# WHY PEOPLE NEED PLANTS

EDITED BY

CARLTON WOOD & NICOLETTE HABGOOD

Kew Publishing
Royal Botanic Gardens, Kew

First published in 2010 by the Royal Botanic Gardens, Kew, Richmond, Surrey TW9 3AB, UK; www.kew.org.
In association with The Open University, Walton Hall, Milton Keynes MK7 6AA, UK; www.open.ac.uk.

ISBN 978 1 84246 425 0

Rachel Parker, Mary Smith, Richard Walden, Patricia Wiltshire and Carlton Wood authored significant sections of this book. The many additional contributors to this work are listed in the acknowledgements (p. 183).

This book forms part of the Open University course *S173 Plants and people*. Details of this and other Open University courses can be obtained from the Student Registration and Enquiry Service, The Open University, PO Box 197, Milton Keynes MK7 6BJ, UK (tel. +44 (0)845 300 60 90, email general-enquiries@open.ac.uk).

British Library Cataloguing in Publication Data
A catalogue record for this book is available from the British Library.

Printed and bound in Italy by Printer Trento

For information or to purchase all Kew titles please visit
www.kewbooks.com or email publishing@kew.org

Kew's mission is to inspire and deliver science-based plant conservation worldwide, enhancing the quality of life.

All proceeds go to support Kew's work in saving the world's plants for life.

# Contents


| | | | | |
|---|---|---|---|---|
| | | Chapter 1 | **Introduction** | 5 |
| I | Uses of plants | Chapter 2 | **Food crops** | 13 |
| | | Chapter 3 | **Wood, fibre and starch crops** | 29 |
| | | Chapter 4 | **Biofuels** | 41 |
| | | Chapter 5 | **Plants in crime** | 51 |
| II | Plants and health | Chapter 6 | **Plants for nutrition and well-being** | 63 |
| | | Chapter 7 | **Medicinal plants** | 75 |
| | | Chapter 8 | **Drink and drugs** | 89 |
| III | Modern techniques in plant biology | Chapter 9 | **Micropropagation** | 101 |
| | | Chapter 10 | **Genetically modified plants** | 111 |
| | | Chapter 11 | **Natural plant protection** | 127 |
| IV | Plants and the planet | Chapter 12 | **The impact of humankind on the planet** | 137 |
| | | Chapter 13 | **Conservation** | 149 |
| | | Chapter 14 | **Plant collecting and trading** | 161 |
| | | Chapter 15 | **Plants and the future** | 173 |
| | | | **Photography credits** | 179 |
| | | | **Acknowledgements** | 183 |
| | | | **Index** | 184 |

# 1 Introduction

If you look up from the pages of this book and glance around you, the chances are that you will see the distinctive green of a living plant. Even if you can't see any flower beds or trees, you may see parts of plants that have been recently harvested (vegetables, rice and other grains, herbs or cut flowers) or perhaps products that were made from plants (such as bread, pasta, tea and coffee). Plants may well have been used to make some of your clothing, and there is likely to be wood in any furniture that you can see. Even if we are not consciously aware of it, plants play a wide range of significant roles in our lives.

At a fundamental level, plants are an integral part of life on Earth: not least because humans and most other animals depend on them for nourishment and the oxygen in the air we breathe.

**Figure 1.1**
A selection of products that originate from plants.

## Why people need plants

- Plants are an indispensable part of the global biosphere; they regulate the atmosphere and provide the first step in the food chain for almost all animals.
- Plants are the foundation of the economy, both nationally and globally.
- Many plant products have important practical uses, such as providing us with building materials, fibres, fuels and medicines.
- Growing and harvesting crops, and working with plant material, provides employment for millions of people.
- Plants offer many intangible benefits including the beauty of their form, colour and fragrance, and the satisfaction we can derive from nurturing them for pleasure.

## What is a plant?

The key thing that separates plants from other living things is that they can capture sunlight to power the process of photosynthesis. This amazing chemical process captures gaseous carbon dioxide that would otherwise accumulate in the atmosphere, hastening global warming, and combines it with water to make simple sugars and oxygen. A plant produces more oxygen than it uses, so we benefit from the excess. The simple sugars made by photosynthesis are turned into an extraordinary range of compounds, which are the raw materials that make up the plant itself.

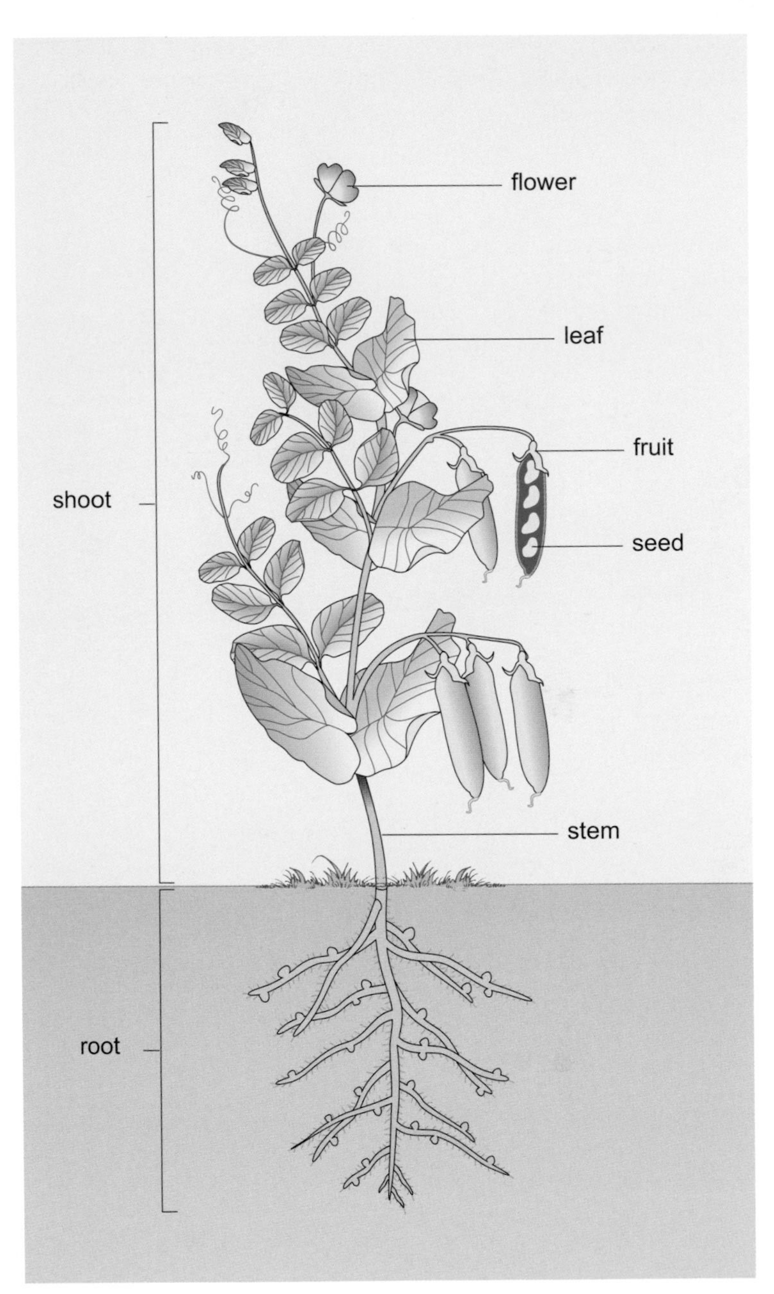


**Figure 1.2**

Plants have roots, leaves and stems.

Plants come in many forms, mosses and liverworts are plants, so are ferns, but this book deals with the flowering plants that are instantly recognisable: the trees, shrubs, grasses and herbaceous plants found in our gardens, parks and fields. These plants come in many different shapes and sizes but they all share a basic set of structures. Roots anchor plants within the ground, enabling them to take up water and minerals from the soil and to interact with essential fungi. A plant's stem contains tissues which transport water and nutrients around the plant. Leaves, which are typically green, are the primary site of photosynthesis. Flowers are the reproductive parts of a plant. Seeds are produced after flowering and often contain a food reserve to nourish the young plant when the seed germinates and begins to grow. Seeds may be contained within a fruit, for example in tomatoes, peas and apples. Plants also store food in swollen leaves, roots or stems and these help the plant survive drought or freezing. We use these food stores ourselves, for example potatoes are swollen underground stems called tubers, carrots are swollen roots and onions are swollen leaf bases. The term 'shoot' can refer to all the green parts of the plant above the ground, but often it refers to the growing tip of the plant from which the new stem and leaves are produced.

## What are plants made of?

All parts of a plant are made from cells, little 'bags' containing everything a plant needs to live, grow and reproduce. They are, in essence, the building blocks from which a plant is constructed.

The cells have thick walls, made from a tough substance called cellulose. Each cell has a nucleus, which contains the DNA needed for cell division and reproduction; the nucleus also gives instructions to the cell to determine which proteins are produced. A vacuole, which stores water and other molecules and provides support for the cell, forms a large part of the cell volume. In the green parts of the plant, the cells contain structures that are responsible for photosynthesis known as chloroplasts, which means 'coloured bodies'. Surrounding the chloroplasts, nucleus and vacuole is a gel-like liquid known as the cytosol.

**Figure 1.3**

(a) Cross-section of part of a maize leaf showing several plant cells together, taken with an electron microscope.

(b) A drawing of one of the cells in (a) with the main features labelled.

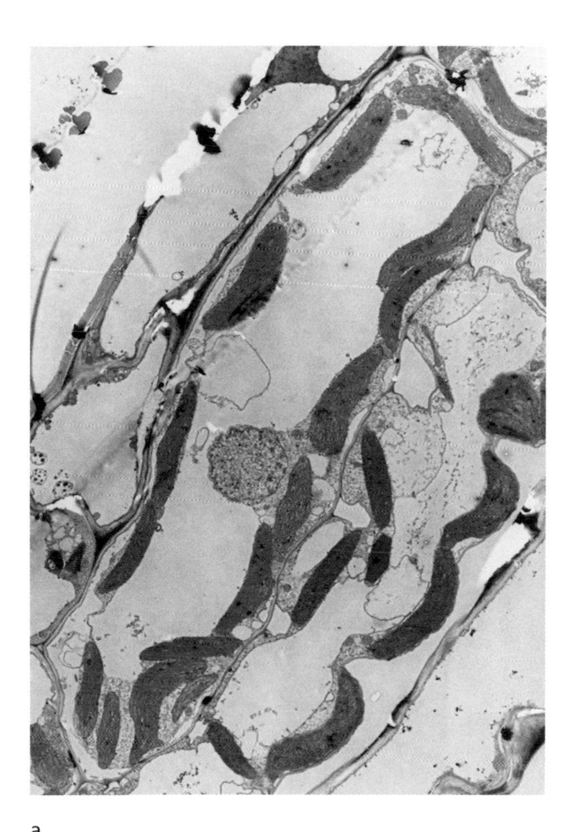

a

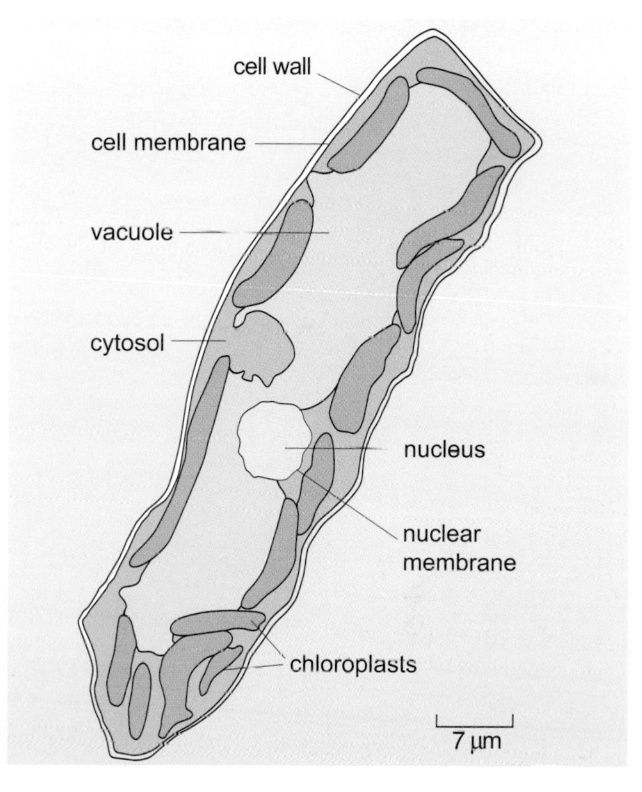


b

The cells within a plant are organised in groups to form the different types of tissue in various parts of the plant. Each tissue has its own function; for example, the tissue in the stem of a flowering plant transfers liquids around the plant and keeps the plant upright; so the cells in the stem are elongated and organised to give strong, waterproofed, elongate tissues.

## Photosynthesis in plants

Photosynthesis is a complex process that involves the plant using energy from sunlight to combine carbon dioxide gas with water to make sugar. The reaction can be expressed in words as follows:

$$\text{water + carbon dioxide} \longrightarrow \text{sugar + oxygen}$$

The sugar is typically in the form of glucose, but it could be other kinds of sugar, or indeed sugar alcohols. The oxygen that is a by-product of the reaction is essential for all plant and animal life.

Photosynthesis occurs inside chloroplasts, which are found in the leaves (and other green parts) of the plant. Chloroplasts contain molecules of chlorophyll, a green pigment that captures energy from sunlight and harnesses it to enable the sugar to be produced. Chemically, plant tissue is mostly made up of compounds containing carbon (known as organic compounds), which are made by a very complex set of chemical reactions from the sugar produced during photosynthesis. Around 44% of the dry weight (that is, weight excluding water) of a living plant is carbon, captured from the air during photosynthesis.

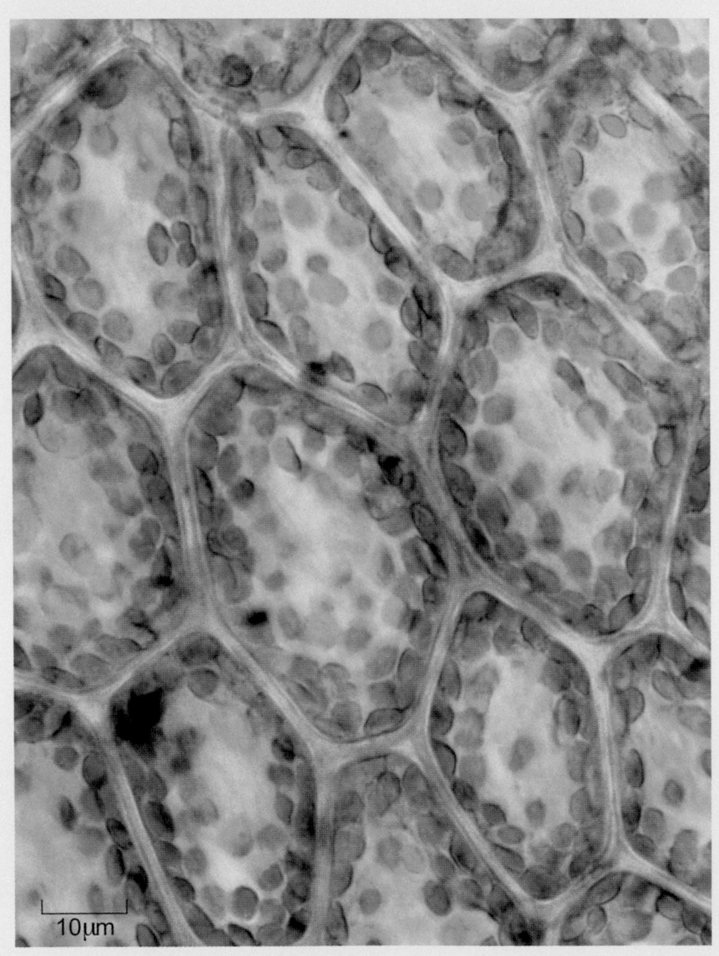


**Figure 1.4**

Cross-section through a leaf observed with a light microscope. Seven complete cells can be seen, each of which contains many round, dark green structures known as chloroplasts.

## Why are plants green?

An isolated extract of chlorophyll is dark green in colour. Sunlight (white light) comprises all the colours of the rainbow. Chlorophyll absorbs much of the blue and red light that falls on it, but most of the green light is reflected and so can be detected by our eyes. This is why chlorophyll itself and the vast majority of plant leaves appear green. Interestingly, if the photosynthetic pigments absorbed all the colours of light the leaves would appear black and, as black surfaces absorb heat particularly efficiently, plants would probably overheat.

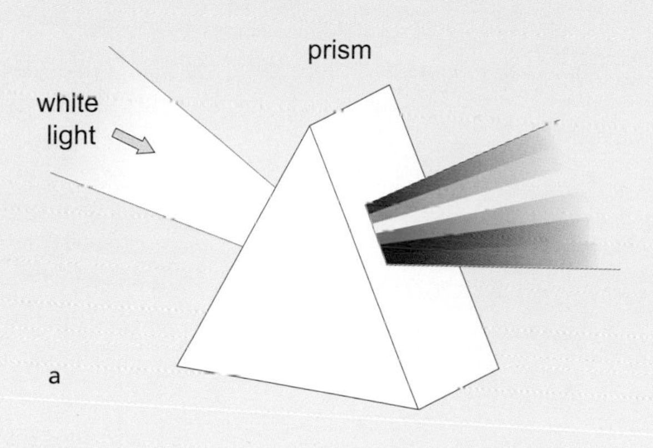


suspension of chloroplasts

white light

green light

b

**Figure 1.5**

(a) The different colours in white light can be seen when light is shone through a prism.

(b) When white light is shone through a suspension of chloroplasts, the chloroplasts absorb all the colours with the exception of the green light.

Plant tissues are a useful source of food because either they still contain the sugars produced by photosynthesis or they contain the sugars in a stored form, as starch or oil. The total annual tonnage of plant material harvested worldwide is staggering: in 2007 just over seven thousand million (7,000,000,000) tonnes were produced.

Luckily for us, plants are amazingly productive. It has been estimated that across the world, in one single second, growing plants capture more than one hundred thousand million (100,000,000,000) kilograms of carbon, which they use to build sugars. Thus photosynthesis enables plants to make enough sugar for every human in the world to have 15 kg each, every second!

## The blue-green agave: a prodigious producer of sugar

The blue-green agave (*Agave tequilana*) flourishes in the hot, dry areas surrounding the city of Guadalajara in Mexico. The distilled product obtained from the sugary juice of the modified stem (or piña) is the alcoholic drink known as tequila. However, it will come as no surprise that agave does not produce the sugary juice for human benefit.

A single plant will grow for around 10 years, producing complex sugars called fructans, which are stored within the piña. After a decade, the piña can weigh as much as 80 kg. When the time comes for the plant to flower, which it does only once in its lifetime, all of the sugar reserves in the piña are mobilised and, within about 30 days, the plant grows a flowering stalk which will typically be several metres tall and tens of centimetres in diameter.

**Figure 1.6**
The blue-green agave plant (*Agave tequilana*) and tequila, the alcoholic beverage which is produced from it. The piña, which contains the stored sugar, can be thought of as the 'stem' of the plant and is the tissue to which the bases of the leaves are attached.

### How are plants named?

Many plants are known by more than one name, even within the same country; for example, goosegrass, which grows wild in the UK, is also known as cleavers, beggar lice, hariff, gripgrass and catchweed. Even more confusingly, sometimes the same name is used for more than one plant; for example, in Scotland the plant known as a bluebell is what is known as a harebell in the rest of Britain. To avoid such confusion, scientists worldwide give every type of plant a unique scientific name.

The scientific names are based on Latin and are in two parts. The first part of the name refers to the plant's genus (the plural is 'genera'), which can be thought of as a group of related species. The second part is the specific epithet and is frequently an adjective that helps describe the species to which it applies. Together, the genus and the specific epithet constitute the name of the species. Scientists define a species as a group of organisms in which the members are able to breed together to produce viable offspring.

In the example of *Agave tequilana*, Agave is the name of the genus, tequilana is the specific epithet and *Agave tequilana* is the species name. Latin names are usually written in italics or are underlined. After the species name, you may sometimes see a shortened version of the name of the person who first described the species. In the case of *Agave tequilana* this person was a German botanist named Franz Weber, so the full botanical name of the blue-green agave becomes *Agave tequilana* Weber. In the case of goosegrass, the species name is *Galium aparine* L. The letter 'L' here is particularly significant as it refers to the Swedish botanist Carl Linnaeus, who devised this method of scientifically naming species.

All organisms can be classified into broader groups above the level of the species or genus. In terms of increasingly broad groupings, the five additional classification levels are: family, order, class, division and finally kingdom. Plants belong to the kingdom Plantae; humans (*Homo sapiens*) belong to the kingdom Animalia. The science of classifying organisms into groups that reflect how they are related to one another is known as taxonomy, and is very important in helping us identify plants correctly and so distinguish the ones that are useful to us.

| level | scientific name | common name |
|---|---|---|
| kingdom | Plantae | plants |
| division | Magnoliophyta | |
| class | Liliopsida | |
| order | Liliales | lilies |
| family | Agavaceae | century-plant family |
| genus | *Agave* | |
| species | *Agave tequilana* | blue-green agave |

**Figure 1.7**
The classification levels for *Agave tequilana*.

## Overview

Humans exploit the extraordinary productivity of plants not only to obtain energy in the form of food and drink but also to use them as fuels, which when burnt release energy for heat and power. This is explored in **Uses of plants**; whilst **Modern techniques in plant biology** examines the methods farmers and scientists are using to tackle the challenge of feeding a growing world population.

For millennia, people have also used plants to improve their lives by providing remedies to alleviate the symptoms of illness and injury, and to obtain stimulating drinks and drugs that induce feelings of well-being. These traditional uses of plants are discussed in **Plants and health**, which begins with a basic outline of the modern principles of nutrition that includes the central role played by plants.

The final part of the book, **Plants and the planet**, shows how the exploitation of plants has affected the planet and looks at how we are striving to meet the challenges of maintaining plant diversity and ensuring that our use of these invaluable resources is sustainable.

# 2 Food crops

A crop plant is one that is cultivated to provide people with a useful product. The commercially valuable product may be the plant itself, for example leaves, fruit, seed or tuber for food; stems for wood or fibre; or flowers for their beauty and perfume. Or it may be an extract or by-product, for example oil from rape seed or olives, or the bioethanol fuel produced by fermenting sugar-rich plants such as sugar beet.

## 2.1 Survival before farming

For more than three-quarters of our five hundred thousand year history, humans lived in small groups of hunter-gatherers, who moved through the landscape collecting foodstuffs as and when they required them. When demand for these resources exceeded the capacity of the ecosystem to supply them, a group simply moved on to another site. Left untouched, the ecosystem resources in the area that had been exploited would usually recover to previous levels.

When supplies of food plants were plentiful, food gathering might only take about three days a week, and living was relatively easy. However, this way of life was very dependent on what nature could provide, which would vary from year to year, and with the seasons. Supplies could become exhausted, forcing a group to move on.

Human population levels were controlled by the availability of renewable resources: plants and animals for food, and wood for fuel. In difficult periods, social pressures resulted in geronticide and infanticide (the killing of the oldest and youngest members of the group); these are harsh (but evolutionarily sound) ways of reducing demand for resources. Over time, people developed not only tools to make hunting and gathering more effective, but also clothing for warmth, so reducing the amount of both fuel to be collected and food that needed to be consumed. As more efficient ways of exploiting the landscape led to a growing population, however, so demands for natural resources increased.

## 2.2 The first food crops

Agriculture began to develop when human demand for food exceeded the supply available from natural vegetation. To survive, people had to settle in one place and cultivate the land. This meant investing more time and energy in food production, as cultivation requires seeds to be sown, plants to be cared for during the growing season and the crop to be harvested.

Hunter-gatherers may have taken up farming as a result of changes in

**Figure 2.1**

In the Fertile Crescent 11,000 years ago, the demand for food from the growing human population began to exceed that which could be obtained by foraging. Skills such as the selection of plant species for crops and the development of husbandry and harvesting techniques allowed supply to meet demand. Agriculture was adopted on a global scale in a relatively short time. But bad weather, resulting in poor crops, meant that, for individual communities, famine was still never far away.

environmental conditions that came with the ending of the last glacial period, around 10–15,000 years ago. Agriculture arose in various geographical locations over a period of time. It is known that extensive wheat cultivation was taking place about 11,000 years ago in an area known as the Fertile Crescent, a curved area that stretched from the northern end of the Persian Gulf up to the eastern end of the Mediterranean and incorporated the ancient kingdoms of Mesopotamia and Assyria. By 7,000 years ago, rice was being cultivated on the banks of the Yangtze River in China.

## 2.3 Crop improvement

Cultivation offers potential to increase the amount of food available. Farmers wished to make the crops more productive, and to achieve this they added plant selection to their farming skills. For example, the wild relatives of what would become modern-day wheat and rice varieties had a mechanism that allowed their mature seed heads to 'shatter', or shed seed, so their seed was readily released and dispersed by the wind. This trait meant that when these wild varieties were cultivated, a significant amount of grain could be lost during harvest. Very early in wheat and rice domestication, certain plants were observed to have grains that did not shed their seed and, by accident or design, these plants were selected and bred from, resulting in improved crop yields.

### What are seeds?

Seeds are nature's way of allowing plants to survive when conditions are not conducive to plant growth. Seeds contain a dormant embryo which, in favourable conditions, will begin to grow (germinate). They also contain storage products, such as starch, oil or even protein, which are used as an energy source to support the growth of the embryo until it can photosynthesise sufficiently to sustain itself. The storage products are contained within the cotyledons, which develop into the first leaves of the seedling (the term 'cotyledon' means 'seed leaf').

Grasses and cereals are described as monocotyledonous plants (often called 'monocots'), which means they have only one cotyledon, leading to one seedling leaf (mono means 'one'). Monocots generally have long, thin leaves and parallel veins. Dicotyledonous plants ('eudicots') are characterised by having two cotyledons and seedlings with two leaves.

An outer seed coat, called the testa, surrounds the embryo. The seeds are often protected by other plant tissues that contribute to the structure that botanists call a fruit. This may be something a cook would also call a fruit, like an apple, but often a fruit to a botanist is commonly known by another name. For example, tomatoes, peas and cucumbers are all technically fruits. In some seeds, from which cereals and grasses develop, there is extra tissue, the pericarp, fused to the testa. Hence a grass or cereal grain is in fact a fruit, albeit a dry one with one seed! The term 'grain' is often used to refer to the seeds and fruits of a range of economically important crops.

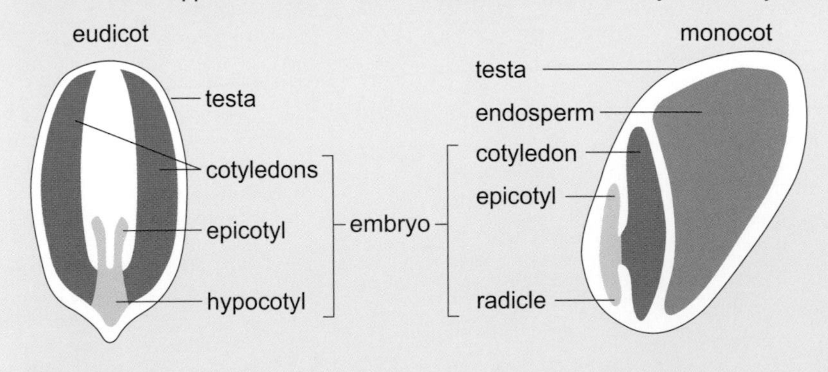


**Figure 2.2**
The structure of a eudicot and a moncot seed.

Seeds, if stored in suitably dry conditions, can remain viable for many years without germinating. But under optimal temperature, oxygen and light conditions, the uptake of water stimulates germination. The water releases enzymes from the embryo and these break down and mobilise stored carbohydrates or oils, which will be used to support the growth of the seedling.

## What is selection?

Anyone growing crops likes to ensure that they get as much as possible of the part of the plant they want. Early farmers were no different and, although evidence is hard to come by, it seems likely that they collected seeds from the most productive wild plants in the environment around them, and sowed those.

This is the first step in a selective breeding programme. The next step is to collect and save seed from the first generation of plants cultivated and sow it to produce the second generation, and so on. Whether by accident or design, farmers may select for characteristics that they find useful, so that over many generations, the crops that are grown resemble their wild relatives less and less.

The third step occurs when two closely related plants breed together (or cross-fertilise) to produce offspring with a combination of characteristics that are more desirable than those of either of the parent plants. Modern plant breeding involves careful human manipulation of cross-fertilisation, but it is hard to know whether early farmers caused this to happen deliberately or simply benefitted from natural variation.

In nature, few plants are perfectly suited to our needs, and throughout our agricultural history we have selected plants that have prized characteristics (for example, high yields of fruit or good flavour) and used them in breeding programmes in an attempt to produce plants with the desired attributes. In selecting cereals, ease of threshing (separating the grain from its husk) is a valuable characteristic, whereas in apples, levels of sweetness (determined largely by the content of the sugar fructose) are important.

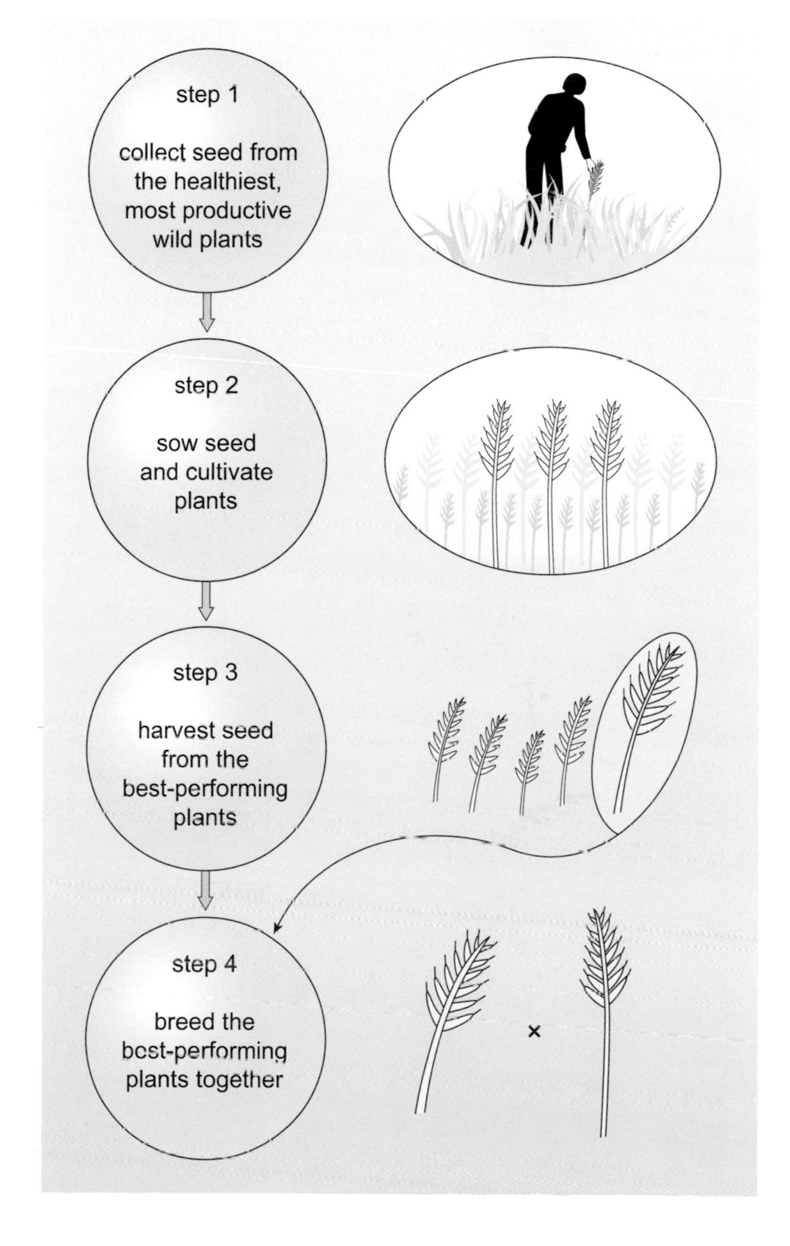


**Figure 2.3**
The stages in a selective breeding programme.

Other characteristics that may be desirable include those which make plants:

- produce more (or fewer) seeds
- produce larger (or smaller) seeds
- produce an end product that can be stored or easily processed
- have more manageable growth characteristics, for example shorter stems
- have reduced seed dormancy and high rates of germination
- have increased resistance to pests and diseases.

The majority of modern crop species have arisen through the selection of particular plants or their hybrids.

As mentioned earlier, plant domestication occurred in a number of geographical locations, and it is often necessary in breeding programmes to go back to the places where the crop plants in question originated, so as to access the range of genetic variability present in their wild relatives. Armed with information about the potential genetic variability of the crop, plant breeders can achieve improvements in crop yield by choosing certain genetic characteristics from those present in the wild relatives, so as better to match the crop to the environment in which it is being grown. At the same time, the agricultural environments in question may be made more uniformly productive through soil cultivation, application of fertilisers and irrigation.

## The origin of domesticated wheat

Through the work of the botanist and geneticist Nikolai Vavilov and others, the geographic locations ('centres of origin') of the wild relatives of the world's cereal crop plants are known with some certainty. Wheat (*Triticum* species) is one of the world's most important food crops. It originated in the eastern Mediterranean/Middle East area, and there is historical evidence to show it was one of the first plant genera to be cultivated and bred over 10,000 years ago. The two early wheat forms at the start of wheat domestication were einkorn wheat and emmer wheat, a low yielding, awned wheat (awns are the 'whiskers' on the developing ear). The first wheat cultivars (species bred specifically for cultivation) spread across various continents during the Neolithic period around 11,000 years ago. Since then, wheat has been bred continuously to give higher yields, better disease resistance and fewer and shorter awns.

a

b

**Figure 2.4**

Historical and scientific evidence has shown that (a) einkorn and (b) emmer wheat are the precursors to domesticated wheat cultivars.

Born in Moscow, Nikolai Vavilov was a botanist and geneticist who undertook over 180 plant-collecting expeditions during the early years of the twentieth century. He proposed that each type of cultivated plant originated from a particular place where the greatest variation in its wild relatives can still be found. Vavilov called these places 'centres of diversity' and claimed they are also centres of origin.

Moreover, Vavilov recognised that these wild relatives represented important, and possibly unique, pools of genetic variability that could be helpful in plant breeding. He initiated efforts to set up secure collections of plants and seeds, including one in St Petersburg with a store of 250,000 seed samples, to preserve them for the future benefit of plant breeders. These efforts continue today at various seed banks set up throughout the world.

**Figure 2.5**

The Russian plant geneticist Nikolai Vavilov 1887–1943, who proposed the concept of centres of origin.

**Figure 2.6**

Centres of origin for a variety of crop plants.

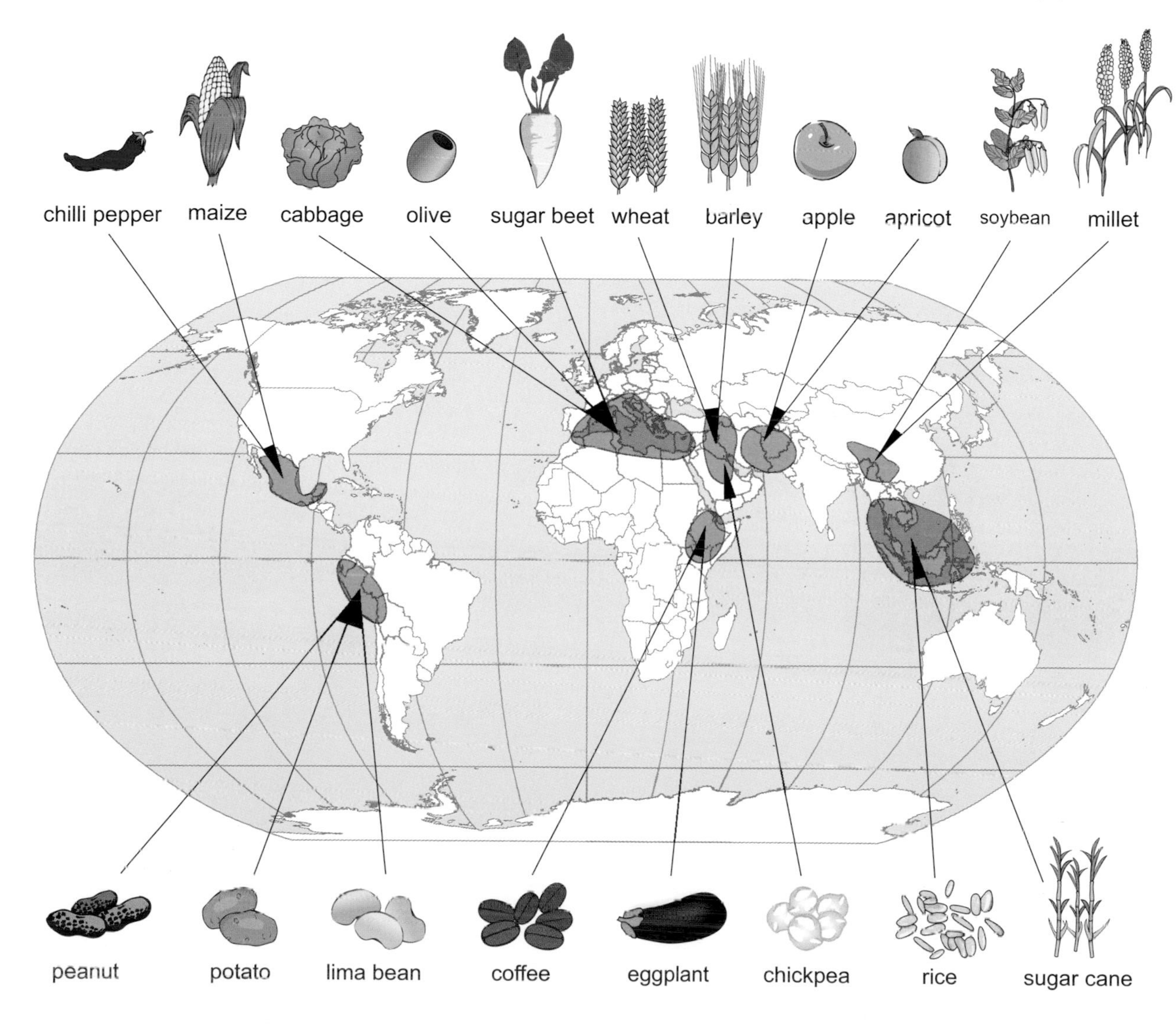

## Modern plant breeding

Today, plant breeders aim to improve crop varieties through cross-fertilisation and subsequent hybridisation, rather than by the selection of wild plants. The two parent plants are chosen for different characteristics that are judged to be useful, and when the parent plants are bred together, or cross-fertilised, these characters may or may not be combined in their offspring.

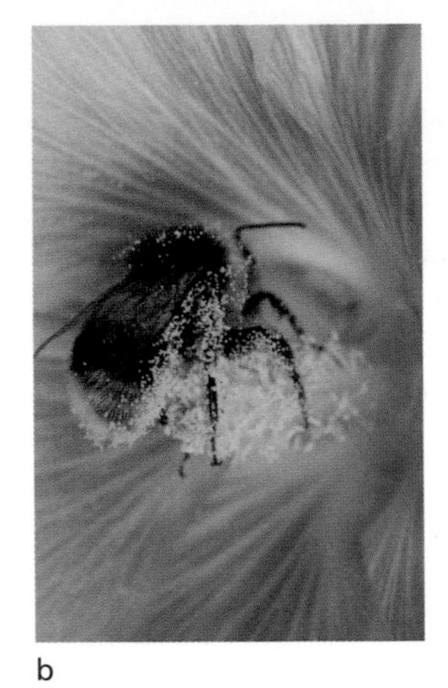

**Figure 2.7**

Cross-fertilisation taking place (a) artificially using a brush to transfer the male pollen to the female parts of the flower and (b) naturally by a bee.

Plant breeding can be a time-consuming process and requires considerable resources. Moreover, a breeder needs skill, detailed knowledge of the crop and its genetics, an eye for useful and undesirable characteristics, good record keeping, and a lot of patience. While the probable outcomes of a breeding programme may be predicted, a large element of luck is involved, as all the best characteristics of the parents (or the ones that are specifically required) are not always passed down to the offspring. The process takes many years and almost every plant breeder can tell a demoralising tale of how seeds or seedlings were lost at a key stage because of bad weather or other unforeseen events, thus negating the results of several years of painstaking work.

Increased yield is an obvious goal for selective breeding, but there can be other benefits too. Imagine walking through a field of wheat on a summer day in twenty-first century Europe: the ripening grain spikes, the seed heads and their associated stems are about a metre, or slightly less, off the ground, just below adult waist height. This wheat is very different from the shoulder-high crops depicted in medieval paintings, and is very much more productive.

**Figure 2.8**

(a) *Harvesting Corn* (1566) by Pieter Brueghel the Elder. (Note that in Europe 'corn' refers to wheat, whereas in the United States it refers to maize.)

(b) A modern-day farmer in a field of wheat just prior to harvest.

The shortness of the 'new' wheat (a characteristic known as 'semi-dwarfing') is an outcome of breeding in the 1950s and 1960s aimed at improving yields. One particular breeding programme resulted in the development of a type of wheat known as Norin 10. This particular cultivar was semi-dwarf but produced large, high-yielding ears; it helped increase the wheat production of developing countries during the Green Revolution, a period between the 1950s and the 1970s when agricultural practices improved dramatically. A dwarf structure means that a variety requires less water and fewer nutrients, and it reduces the chances that the plant will be blown over, thus increasing the efficiency with which it can be harvested. However, the main advantage of the dwarfing characteristic is that it results in the plant distributing proportionately less of the carbon fixed by photosynthesis into leaves and stems, leaving more available for the grain. Newer wheat varieties now contain about 50% of their carbon in grains, compared to only about 20% in older varieties.

Another example of a cultivar whose introduction had far-reaching effects is a rice called IR36, which was produced in the latter half of the Green Revolution. IR36 is semi-dwarf, high yielding and matures rapidly: characteristics that led to it being grown on a huge scale. At one point, 7% (11 million hectares) of the global rice growing area (154 million hectares) was planted with IR36.

## 2.4 Staple food crops

Staple crops can be defined as those that provide the main food of a community. Most staple crops produce grains (seeds and fruits) or tubers which when mature have a relatively low moisture content, which aids storage. Most staple crops are relatively high in starch, an energy-rich carbohydrate derived from the sugars produced during photosynthesis.

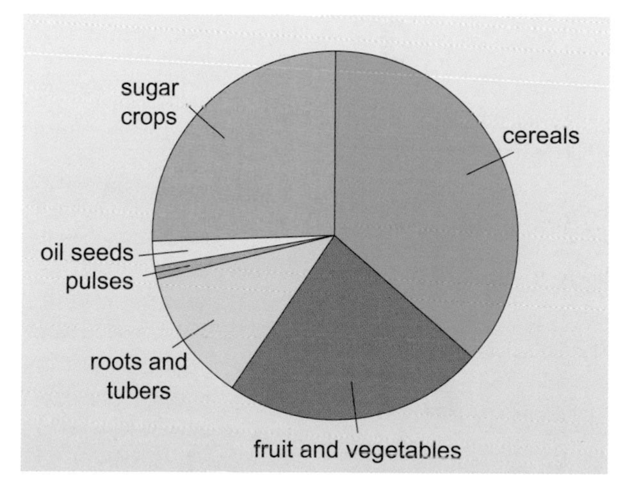


**Figure 2.9**
The relative amounts of different crop types that are harvested annually worldwide. The term 'pulses' refers to crops such as peas and beans, whilst 'cereals' refers to crops such as rice, wheat, maize and sorghum.

### Grains as a food crop

Seeds and fruits of a range of economically important crops are often known as 'grains'. Many different kinds of grains are used as staple foods and some 55% of global cultivated land is dedicated to grain production, with 76% of that being planted with maize, wheat and rice. Grains are often used to feed livestock, and grains and livestock together provide about 80% of the world's food supply in terms of calories consumed.

Whole (complete) grains contain approximately 75% starch, up to 15% protein, about 2% fat and vitamins such as B and E. For maximum nutritional benefit to humans, the grain should be eaten whole. White flour is made from just the endosperm tissue of the grain (Figure 2.2) and contains neither the fibre-rich bran (found in the grain's testa) nor the vitamin-rich germ (the embryo).

Starch from grains is the main source of calorific intake in western diets and contributes up to 80% of calorific intake in developing countries. Grains also contain certain amino acids that can be crucial in helping to provide a balanced diet. The outer layers of grains contain various vitamins, including vitamin $B_1$; indeed, vitamins were discovered following a study on cereal grains.

## Beriberi

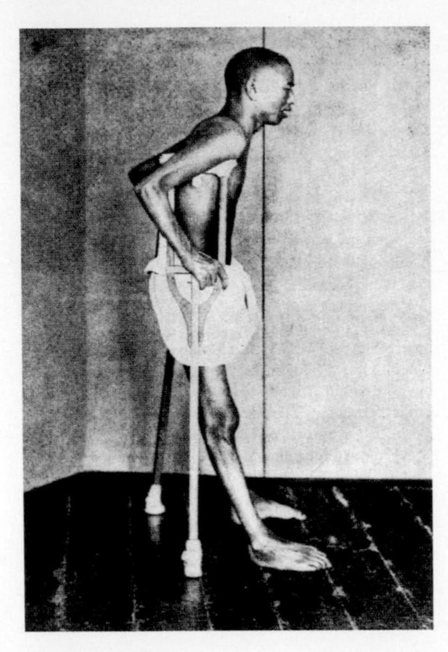

The disease beriberi was first described by the Chinese in the seventh century, when it was observed in populations subsisting on a diet of rice. The number of cases in South-East Asia increased dramatically in the 1870s after steam-driven mills to polish rice grains were introduced from Europe. Polishing completely removes the testa and embryo from rice. Although it was more expensive than conventional brown rice, polished rice became popular because it was considered to have a superior taste.

Christiaan Eijkman, an army surgeon working in the Dutch East Indies, studied beriberi, which had become endemic there. He used chickens because they developed the disease rapidly, and found that they became unwell when fed white polished rice but recovered when fed with uncooked, unpolished brown rice. Polishing of the rice removes the parts of the seed containing vitamin $B_1$, and deficiency of vitamin $B_1$ leads to beriberi. For his work on beriberi, Eijkman shared the 1929 Nobel Prize for Medicine.

**Figure 2.10**

In the nineteenth century, beriberi was common among Asian sailors on long voyages, as well as among army labourers and convicts. The disease manifests itself as weakness and fatigue, and, as it progresses, patients report burning sensations, tingling in the extremities, numbness and eventually paralysis. Ultimately, sufferers may die of heart failure.

## Legumes as a food crop

Peas and beans belong to the legume family, in which seeds are enclosed within pods. They are an important staple crop because the seeds contain not only large amounts of carbohydrate, but also relatively large amounts of protein, up to 25% of the dry weight of a seed. In addition, the kinds of protein contained in pulses complement the kinds found in cereals, so that together they provide the basis of a nutritious diet.

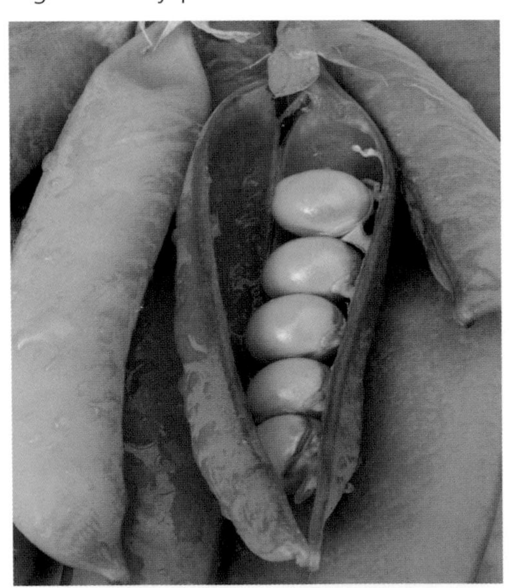

**Figure 2.11**

The seeds of peas and other legumes are enclosed within protective pods.

Legume roots form an association with certain bacteria to capture nitrogen from the atmosphere. These common soil bacteria, *Rhizobium* species, trigger the development of nodules on the roots of legumes. These nodules become the sites of nitrogen fixation, the process whereby nitrogen gas from the atmosphere is converted into nitrogen compounds that can be used by plants, for example nitrates. Nitrogen fixation is essential for all life because fixed nitrogen is required to make the genetic material DNA, as well as amino acids from which proteins are made. In legumes, nitrogen fixation enables the host plant to achieve a high level of amino acids, and therefore proteins, in their seeds.

Some pulses can be eaten fresh, but all have the advantage that they can be dried, stored and subsequently processed. Once the testa is removed, the cotyledons can be separated giving, for example, split peas. The cotyledons can also be ground to form flour, which in turn can be converted into pastes (such as houmus). In addition, pulses can be processed in various ways; for example, soybeans can be fermented to produce soy sauce.

Figure 2.12

Pulses in various shapes and forms: yellow split peas, lentils, dried beans, chickpeas and bottled soy sauce.

Some pulses have to be soaked and boiled prior to consumption because they contain not only potent toxins but also chemicals that reduce the efficiency of digestive enzymes and thus lower the nutritional value. Soaking also reduces the content of sugars that are indigestible to humans but are digested by the bacterial population of the large intestine thereby resulting in flatulence, hence the jokes about eating too many baked beans!

## 2.5 Non-staple food crops

### Crops for oil production

Oil crops are plants that produce oil when their seeds are crushed or pressed. Important oil crops include oilseed rape and linseed in temperate zones, sunflowers and olives in Mediterranean regions, and peanuts and palm oil in more tropical climates. The oil extracted is used for both food and industrial applications.

Figure 2.13

Fields of sunflowers (*Helianthus annuus*) grown for oil production are a common sight in Mediterranean countries, as well as in North America and Russia.

## What are plant oils?

Plant (vegetable) oils are typically extracted from seeds. These oils (liquid fats) are contained within the cotyledon or endosperm tissue, and can make up around 40% of the weight of the seed. Plant oils all have a similar chemical structure in which a molecule called glycerol is attached to three chain-like fatty acid molecules. The glycerol portion is the same in all plant oils, whilst the fatty acid side-chains vary: it is variations in the fatty acid molecules that give each type of oil its own particular characteristics. Each of the fatty acids is a chain typically made up of 16 or 18 carbon atoms to which are bound hydrogen atoms. These fatty acid chains prevent the oil from dissolving (being soluble) in water.

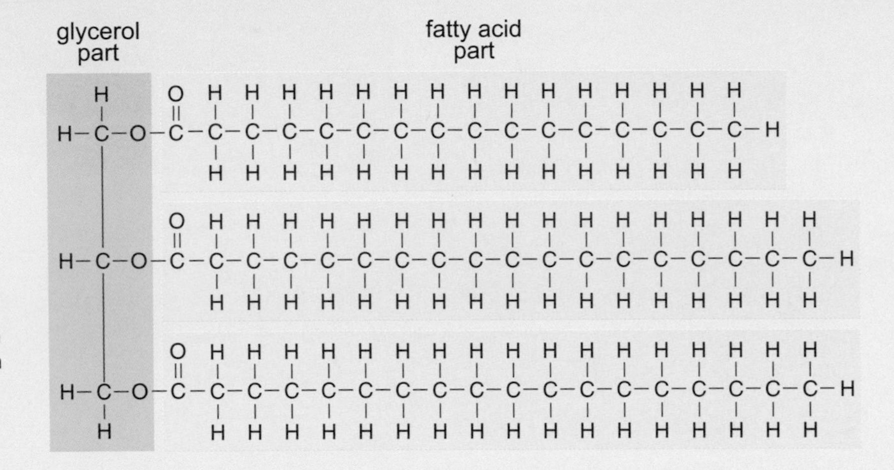


**Figure 2.14**
The basic structure of a plant oil consisting of a glycerol molecule and three fatty acid side-chains.

Oils present within plant tissue will almost always be a mixture, containing fatty acid side-chains of differing lengths and also differing in their degree of saturation, a term that refers to the extent to which the carbon atoms are bonded together with single or double bonds. If there are only single bonds within the fatty acid chains, the oil is referred to as being a saturated oil, and it tends to be solid at room temperature. Cocoa butter contains the saturated fatty acids stearic acid and palmitic acid and these give it the characteristics of a saturated oil. So chocolate is solid at room temperature, but melts in the mouth. Most animal fats are saturated. If the fatty acids contain one or more double bonds between the carbon atoms, then the oil is referred to as being unsaturated, and is more likely to be liquid at room temperature.

**Figure 2.15**
Beans from the cocoa tree provide the main ingredient in chocolate, which contains the saturated fatty acids stearic acid and palmitic acid.

**Figure 2.16**
Olive oil pressed from olives (*Olea europaea*) can contain as much as 80% oleic acid. Oleic acid contains one double bond in its carbon chain and is therefore classed as a mono-unsaturated fatty acid.

**Figure 2.17**

Linoleic acid is found in the oil of a whole array of important commercial plant crops, including sunflowers, maize, safflowers, linseed and walnuts. Linoleic acid has two double bonds in its carbon chain, so it is referred to as being polyunsaturated (poly = many) and oils containing it tend to be liquid at room temperature.

## Essential fatty acids

There are some fatty acids which are essential for human health and nutrition but which cannot be made within the human body and therefore must be obtained from the diet. The two classes of essential fatty acids reflect their chemical structure: omega-3 and omega-6. Linoleic acid (an omega-6) and $\alpha$-linolenic acid (an omega-3) are essential fatty acids which are found in green plants, soybeans, linseed oil, oilseed rape (canola) oil and hemp oil.

**Figure 2.18**

Flax produces linseed oil. This oil has high levels of $\alpha$-linolenic acid, a polyunsaturated essential fatty acid.

## Oilseed rape

Oilseed rape (*Brassica napus*) is grown all over the temperate world, with Canada being the main producer. In Europe, where it originated as an oil crop, rape is important because it is one of only a few oil crops that contain linoleic and $\alpha$-linolenic acids, both essential fatty acids. Cultivation on a large scale is fairly recent in Europe, being introduced in the 1980s when the European Union wanted to decrease its reliance on imported plant oils, so growing oilseed rape was encouraged by the provision of government subsidies. In 2007, farmers in the European Union harvested 20.7 million tonnes of oilseed rape.

**Figure 2.19**

The familiar yellow expanses of fields of oilseed rape.

In addition to the essential fatty acids, oilseed rape also contains high levels of a group of chemicals known as glucosinolates, which are responsible for the characteristically bitter taste of vegetables from the brassica family. Plant breeders have tried to develop types of oilseed rape that contain low levels of glucosinolates, while at the same time having high levels of the beneficial linoleic and α-linolenic acids. In the 1970s a cultivar was developed that had reduced levels of both glucosinolate and another undesirable chemical, erucic acid. This cultivar, known as 'Double Zero' (00), was mainly grown to provide oil for food, and is high in monounsaturated fatty acids, such as oleic acid, and polyunsaturated fatty acids, such as linoleic acid and α-linolenic acid. The nutritional value of the oil is enhanced by the presence of vitamin E and phytosterols, which are cholesterol-like molecules that are thought to act to reduce the harmful effects of cholesterol by reducing its absorption into the intestine.

The oil is released from the rape seeds by crushing them and extracting the oil into a solvent, leaving a protein-rich meal which can be used for animal feed. Rape oil can also be used in the industrial manufacture of biofuels, biodegradable plastics and lubricants. However, compared to cereals, oilseed rape is not a very high yielding crop in terms of seed tonnage produced (2.9 tonnes per hectare, compared to around 10 tonnes per hectare for wheat).

## Health benefits of plant oils

Human diets can be deficient in omega-3 fatty acids, which are easily destroyed during cooking at high temperatures and during food processing. Extracting oils from seeds, nuts or fruits often involves heating to high temperatures, which causes the degradation of proteins and vitamin E. It is important to maintain vitamin E levels, not only for the health benefits of vitamin E itself but also because it helps prevent the destruction of the omega-3 fatty acids. This can be achieved by extracting the oil without heating, a process known as cold pressing, which produces an oil often labelled 'extra virgin'. Both rape seed and olives can be treated this way, to yield cold-pressed virgin oils.

Much has been written about the potential health benefits of the polyunsaturated plant-derived oils, and research is focused on defining exactly what these benefits are. It is known, for example, that the human body converts linoleic acid to gamma-linolenic acid (GLA). This in turn is converted into prostaglandins, natural hormones that have been shown to help reduce blood clots and to control blood sugar levels. As well has helping to lower cholesterol, α-linolenic acid can be converted to a form of prostaglandin that is reputed to have benefits for brain and retinal development, especially in newborn babies. Both linoleic and α-linolenic acids have been reported to have anti-inflammatory and immunological effects, and people suffering from rheumatoid arthritis are often advised to eat seeds and nuts that contain high levels of these acids. Both of these fatty acids are needed for the creation of cell membranes and for the synthesis of important chemicals, known as eicosanoids, which act as signalling molecules between cells within the body.

**Figure 2.20**
The health benefits of high levels of omega-3 fatty acids have become a selling point for cold-pressed oils.

## 2.6 Meeting the challenges of a growing world population

People have grown and harvested plants for more than ten thousand years. Before 1900, famines were relatively common even in Europe and still now famine remains a real threat in many parts of the world.

In the twentieth century, artificial fertilisers were increasingly used to boost crop production to meet demands. This was coupled with improvements in irrigation schemes, providing sufficient water when rainfall was inadequate. Because of changes in agricultural practices, along with plant breeding, famine is now rare in developed countries, and today Europe and the rest of the western world has relatively good 'food security'. However, there are many challenges to overcome if sufficient food is to be produced for the growing populations of developing countries, particularly in Africa and Asia.

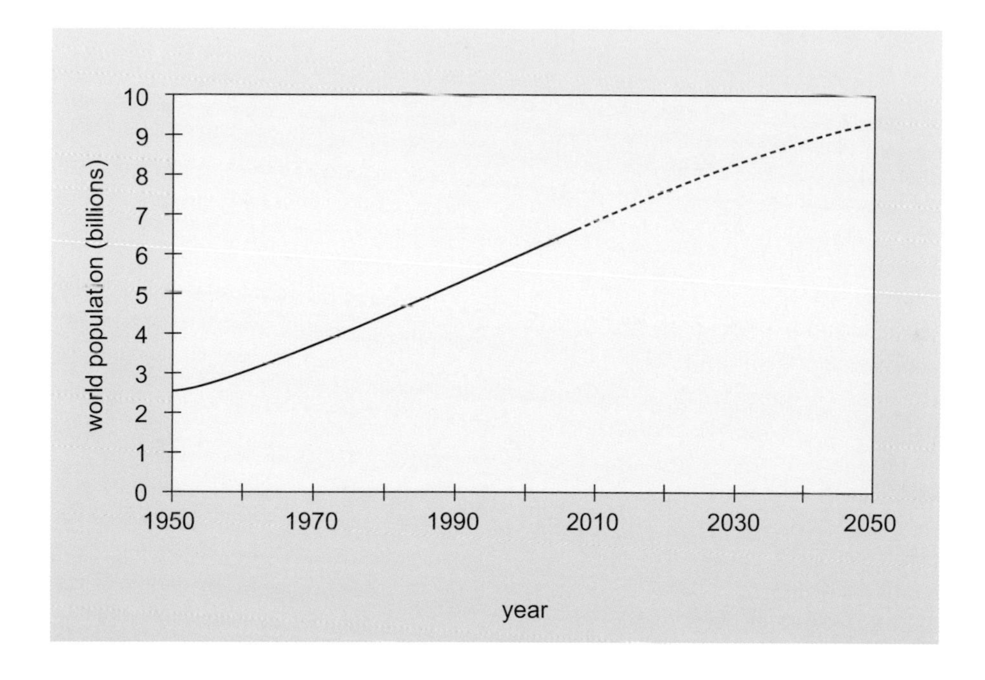


**Figure 2.21**
Increase in world population since 1950 and projected increase to 2050.

Poor harvests and increases in the price of basic foodstuffs often trigger food shortages, and hardly a year goes by without news of a famine somewhere in the world. In six of the years from 1998 to 2007 world grain production failed to match consumption, and by 2009 global reserves of grain had fallen to record lows. In 2008 the price of grain rose to its highest level ever. This had the greatest impact on some of the world's poorest countries, where the proportion of family income spent on food is largest. For example, in Pakistan and Bangladesh, the availability of wheat and rice, respectively, became limited. The situation was exacerbated because both these countries are heavily reliant on imported grain, so when the market price increased, they were particularly badly affected.

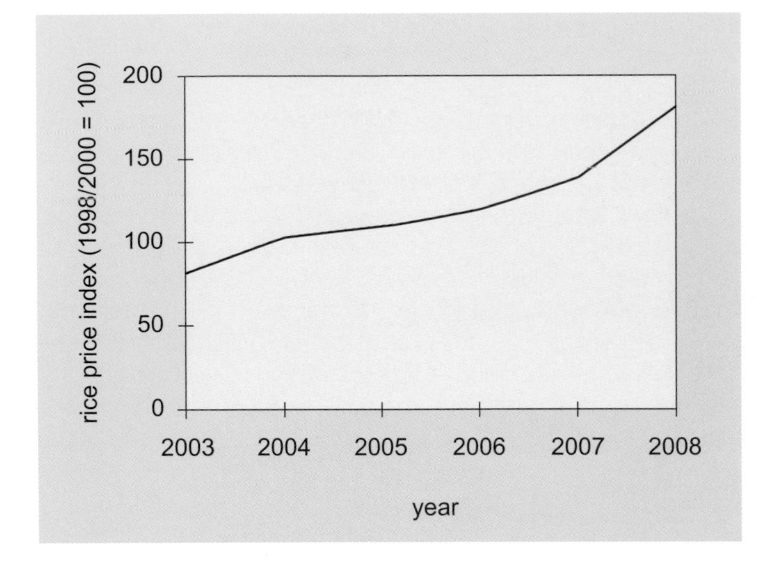


**Figure 2.22**
The rapid increase in the price of rice between 2003 and 2008 was mirrored in the price of other grain crops, causing hardship to populations whose incomes are small.

Conflicts within a country or region can also disrupt the production, supply and distribution of food, resulting in many people becoming severely undernourished. In early 2009, the World Food Programme (an agency of the United Nations) announced that Kenya was in such a situation. Three successive poor harvests (caused by erratic rainfall followed by drought), coupled with political instability after disputed presidential elections, caused an extra two million people within the country to become reliant on the World Food Programme for their survival.

Since the beginning of the Industrial Revolution, ever larger quantities of fossil fuels have been burnt to provide power, and this has significantly increased the concentration of carbon dioxide ($CO_2$) in the atmosphere. Associated with this rise in atmospheric $CO_2$ there has been a rise in global mean surface temperatures along with changes in local climate patterns, affecting both temperature and rainfall; even greater changes are predicted for the future. Furthermore, as global warming continues, we may expect extreme climatic events, resulting in harsh droughts or large-scale flooding of farmland, to become more frequent.

Even using computer modelling, it is difficult to predict exactly how climate change will affect crop productivity and food production as we go forward in the twenty-first century. For countries in regions such as sub-Saharan Africa and parts of Asia, the challenge of feeding their population is likely to become an ever more demanding one. The poorer countries of the world will be the most affected by climate change, although all will be affected eventually.

Rising sea levels, resulting from global warming, are already having an effect in some countries. In Egypt, for example, the fertile soils of the Nile Delta produce more than 60% of the nation's food supply, and are home to two-thirds of the country's rapidly growing population. Much of the delta is already very close to sea level with the effect that the soil is being made saline by sea water flowing into the groundwater. A one-metre rise in the sea level, which many experts think likely within the next 100 years, will result in 20% of the delta being lost. Many other fertile and heavily populated delta regions in the developing world (such as the Ganges–Brahmaputra Delta) are similarly under threat from rising sea level.

Figure 2.23
Drought-cracked soil in a wheat field.

How can such challenges be addressed? It goes without saying that addressing the causes of climate change is high on both global and national political agendas. For a number of years, various organisations across the globe have also been looking at ways in which crop production can be increased in the face of increasing climatic challenges, and at the same time remain sustainable, preserving the environment so that it can continue to produce enough food for future generations.

One of these research organisations is CIMMYT (Centro Internacional de Mejoramiento de Maíz y Trigo), known globally as the International Maize and Wheat Improvement Centre. In the latter half of the twentieth century, CIMMYT and its sister organisations were highly successful in using conventional breeding and improved agricultural practices to increase agricultural yields in those countries where maize, wheat and other crops are grown. CIMMYT's work included the development of fertilisers to increase production, and herbicides and pesticides to prevent crop destruction and help improve crop productivity.

A variety of 'greener' methods are also being promoted for improving harvest yields, with the aim of reducing the use of chemicals (which are expensive and potentially toxic to both humans and the environment) on food crops. There is particular interest in organic foods and methods of production that do not require the use of any synthetic chemicals at all. The challenges are now for crop scientists and farmers to maintain the recent yield increases under ever more difficult circumstances, including a changing climate.

# 3 Wood, fibre and starch crops

In addition to food crops, plants have been exploited by humans for thousands of years: for their ability to clothe us, provide building materials, and form the basis for many natural medicines; for the aesthetic properties of their attractive flowers and leaves; and for numerous other functions. This chapter explores just a few of the ways in which the plant products of wood, cork, fibre and starch contribute beneficially to human life.

In order to understand how plant products are useful to humans, it is helpful to know a little about the basic properties of plants that give rise to particular products.

## How do plants grow and remain upright?

Plants are living organisms that require a supply of water and nutrients to grow, and light to allow them to photosynthesise and grow. Water is taken up by the roots and then transported around the plant through the xylem tissue. Sugars, which are the product of photosynthesis, provide the 'food' or energy supply for the plant and are transported from the leaves to the rest of the plant in the phloem.

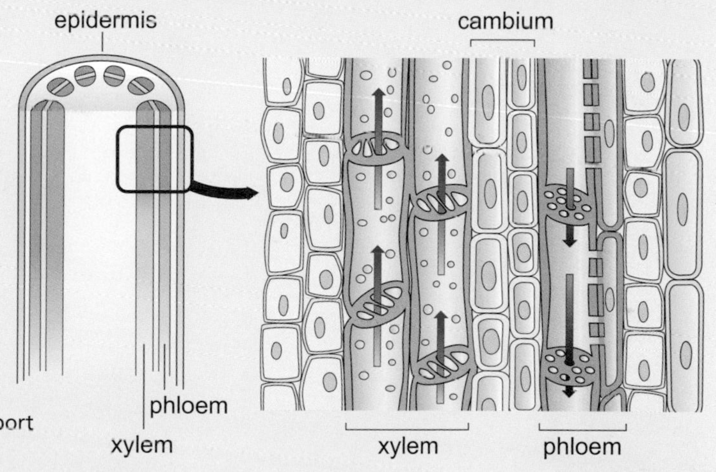


**Figure 3.1**
Typical structure of a plant stem showing the relative positions of the main transport tissues, xylem and phloem.

The xylem and phloem are supported by strengthening tissue called sclerenchyma, which includes specialised cells called sclereids. The cells that make up both xylem and sclerenchyma are characterised by having very thick walls, and, when mature, they contain large amounts of lignin, which strengthens cells and waterproofs them. Tissue containing sclereid cells supports plants by providing rigidity, often in a longitudinal direction, and such tissue is frequently the basis of plant fibres.

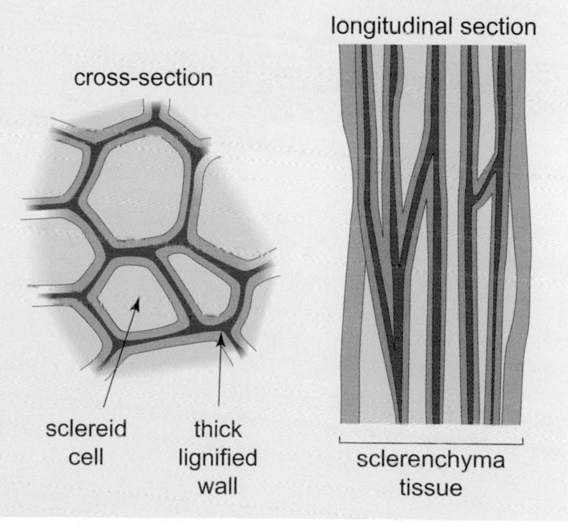


**Figure 3.2**
Schematic cross-section through a stem showing sclereid cells, which are characterised by having thick, strong walls. The way they form sclerenchyma tissue gives rigidity to the plant.

# 3.1 Wood

## What is wood?

Wood is one of the most important and widespread plant products, but its production is quite complicated. In a tree, the water-transporting tissue, the xylem, is in a ring around the woody part of the stem or trunk. Each year a new xylem ring is formed around the outside of the old ring and becomes the tissue that transports the water for that particular year of growth. The previous year's xylem matures, and over a few years dies as it becomes impregnated with various chemicals, including lignin and suberin, making it resistant to decay; eventually it forms the heartwood of the tree.

The rings of xylem become evident when a tree is felled and these annual growth rings can be counted to give the age of the tree. The thickness of the rings gives an indication of the growth rate of the tree through time and can also give clues to past environmental conditions and/or the health of the tree. For example, thicker rings sometimes indicate more rainfall, whereas thinner rings can indicate drought conditions.

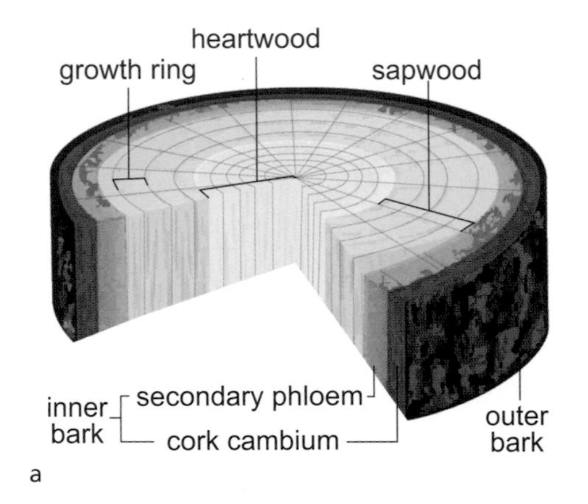

**Figure 3.3**

(a) Labelled cross-section of a tree trunk. The living sapwood is around the edge and contains the functional (current year's) xylem. The bark on the outside of the trunk consists of two layers: an inner layer, which is alive, and an outer layer, which is comprised of dead cells.

(b) The annual growth rings are evident in this cross-section of a larch trunk.

**Figure 3.4**

The heartwood can be completely decayed away but the tree can still survive as long as the sapwood is healthy and transporting water and sugars around the tree via the xylem and phloem, as is evidenced by ancient hollow trees.

There are various different uses for wood and many of them relate to the way in which the wood cell fibres are orientated, giving the so-called grain of the wood. All wood carvers know that it is important to work 'with the grain' of the wood, as working against the grain means that their carvings are not likely to last because the wood will warp or split. The directionality of growth also confers additional properties to wood, such as being able to bend and be flexible whilst still maintaining strength; this is a key reason why wood makes a good building material.

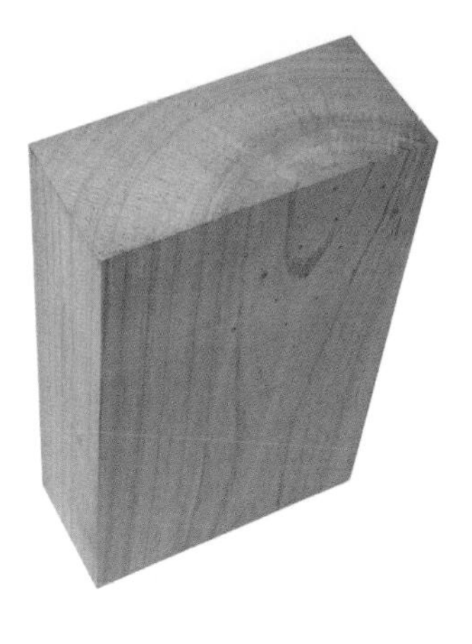

**Figure 3.5**
A block of wood showing the grain, or directionality of its growth.

## Uses of wood

Using satellite-imaging equipment, it has been calculated that there are around 400 billion trees on Earth: this equates to around 60 trees for every human being on the planet. Trees are found on every continent, bar the Antarctic, and because they are so widespread and accessible, humans have used them in many ways throughout history. One of the oldest uses of wood is as a fuel, and this is covered in Chapter 4.

The ubiquitous nature of trees, combined with the fact that wood is relatively easy to work with simple tools, has facilitated the use of wood as a building material since the Neolithic period (between 4,000 and 11,000 years ago) when humans first built shelters. Wood is particularly suitable as a building material, for timber frames and cladding for walls, for floors and for roofs of houses and shelters because it is:

- Insulating: wood conducts heat relatively poorly because of the very small air pockets contained within its cellular structure and so it has excellent insulating properties. It also tends not to expand and contract significantly with changes in temperature.
- Flexible: wood can be bent and still maintain its strength because it is composed of cells, and hence tissues, which are elongated in one direction.
- Light but strong: weight for weight, wood is stronger than reinforced concrete and this is due to the open cellular structure of wood, combined with the presence of lignin in the xylem.

a

b

**Figure 3.6**
Wood is an essential building material for housing across the world, as shown by (a) this wood and mud hut near the Mwalungane Reserve, Kenya and (b) these roof structures being built in Arizona, USA.

a

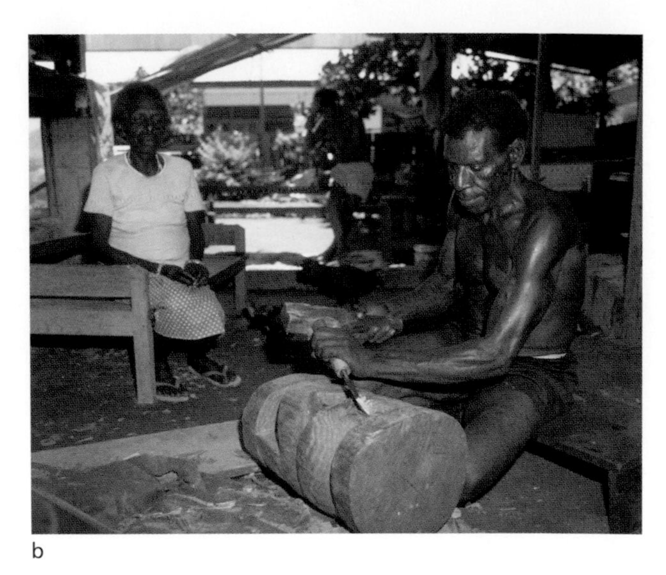

b

**Figure 3.7**

(a) Wood carvings are symbols of power, spiritual life, status and fertility across the world. In the past, totem poles were carved by indigenous people of the north-west coast of North America. The totems, meaning 'his kinship group', were carved out of large trunks of western red cedar (*Thuja plicata*) or other conifers to recount events, tell stories and celebrate cultural beliefs. This totem pole is in the Rocky Mountains, British Columbia, Canada.

(b) Wood carving in Irian Jaya, Indonesia.

In industrialised countries, the demand for cheap, light building materials has led to the development of engineered wood products such as chipboard and medium density fibreboard (MDF). These are composed of very small pieces of wood that are glued together, to give large sheets of wood-based products that are cheaper than natural wood and often have improved properties of dimensional stability.

Before humans developed the technology to work metals, wood was the preferred material in which to carve shapes, patterns or figures. Softwoods, such as pine, are easiest to carve but the wood most commonly used for carving is probably lime or linden (*Tilia*) because it is easy to carve and has a clearly developed straight grain, which aids carving.

Wood can only be considered a renewable resource if trees are replanted at the same rate as they are harvested. Several organisations including the Sustainable Forestry Initiative, the American Tree Farm System and the Forest Stewardship Council have sought to ensure that wood is harvested throughout the world on a renewable basis.

**Figure 3.8**

Over-exploitation of trees has led to many tree species being classified as vulnerable or potentially endangered, including the big-leaved mahogany (*Swietenia macrophylla*) (shown here). It is estimated that this species is being felled and exported from Central America at a rate of 120,000 $m^3$ (90,000 tonnes) a year. The Food and Agriculture Organisation (FAO) of the United Nations estimate that 25% of global carbon emissions are due to the carbon dioxide released when wood from trees is burned following deforestation.

## Forest Stewardship Council (FSC)

The Forest Stewardship Council (FSC) is a worldwide, non-profit-making, non-governmental organisation which promotes the responsible management of the world's forests. It is thought that in some countries 80% of timber exports are from illegal sources that are obtained indiscriminately. The FSC sets standards, maintains trademark assurance and accredits businesses and organisations wishing to supply and manage timber production in a sustainable manner. The FSC certification trademark is an internationally recognised logo and is one of the most important initiatives in promoting a sustainable industry. The trademark assures consumers that the wood they are buying and using comes from forests that are managed to meet social, economic and ecological needs.

© 1996 FSC A.C.

**Figure 3.9**

The FSC logo assures consumers that the timber used in the certified product has been obtained from a managed forest. The organisation also promotes sustainable tree production, through sponsored tree nurseries, and champions human rights, for example by promoting the use of safety equipment for forest workers.

# 3.2 Cork

## What is cork?

Cork is the ultimate renewable resource. Anatomically, it is one of the outer dead layers around a tree trunk, but commercial cork is taken specifically from the cork oak (*Quercus suber*). This tree is native to western Mediterranean countries, and it requires low annual rainfall, mild winters and plenty of sun.

The commercial uses of cork are very diverse and range from the traditional use, recorded by the ancient Egyptians and Greeks, as bottle stops in the wine, cider and cooking oil industries through to its use as a flooring material and for industrial functions. Many of the properties of cork are due to its structure, which resembles that of a honeycomb. Each cell within the 'honeycomb' contains air and suberin, a natural wax-like substance. These characteristics make cork relatively inert, light, waterproof and able to expand and reform itself in different temperature conditions. The suberin also makes cork resistant to rotting. The large air spaces within the cells enable cork to act as a thermal barrier and provide cushioning properties, which have led to a range of industrial uses, such as in expansion joints and as engine gaskets.

**Figure 3.10**

Harvesting cork from a young tree. Removing the outer bark (cork) from the cork oak (*Quercus suber*) without affecting the health of the tree is a skilled task. The cork layer overlies a living layer of tissue and this must not be damaged during the removal of the cork.

## Cork, TCA and the tainting of wine

The best-known use of cork is for producing 'corks' for sealing wine bottles. In recent years plastic 'corks' and screw caps have gained in popularity in the wine industry, as they are said to be as effective as cork at sealing bottles without the possibility of tainting the wine. One of the reasons that some wines 'go off' or are tainted within the bottle is thought to be due to the presence of a chemical called trichloroanisole (TCA) in the cork. When this chemical is present in wine, the wine smells unpleasant. Most people are able to smell the presence of TCA at levels of a few parts per billion. The wine industry is looking to see if TCA can be removed from cork, but at present this is difficult to achieve on a consistent basis.

Cork is a sustainable renewable resource as the trees are not cut down during harvesting and will go on to produce additional cork in future years. Cork oak forests also maintain biodiversity by providing a habitat for some of Europe's rarest wildlife including the Iberian Lynx, which is at the brink of extinction.

The Portuguese cork industry is a good example of sustainable agroforestry where the local people utilise their natural resources for economic gain without destroying the environment in which they live. In addition to the production of cork, the cork oak's leaves are used for animal fodder, fertilisers and tannin preservatives, whilst the wood of old or diseased trees is used for firewood and charcoal. In south-west Spain, pigs are allowed to forage for the fallen cork oak acorns, lending a unique flavour to the Serrano ham produced in the region.

Strict laws are in place to maintain the cork oak forests. In Portugal, for example, a tree can have its cork stripped only once every nine years, to allow time for the cork to grow back. Each tree harvested yields enough cork to produce around 4,000 biodegradable bottle corks. With increasing use of metal screwcaps and plastic 'corks', the cork oak industry, local economies and biodiversity could be seriously threatened.

## 3.3 Fibre

### What is plant fibre?

Plant fibre is difficult to define. Most people will be aware of dietary fibre, the indigestible part of fruit and vegetables that is considered to be roughage within the diet. Technically, for most fibres the term refers to the elongated strengthening sclereid cells found around phloem tissue in the stems of plants, as is the case with jute which comes from the stems of *Corchorus* species, although it should be noted that wood also contains fibres. Fibres can also be found in the leaves of some plants (such as sisal) and around the fruit and its coat, for example in coconut (*Cocos nucifera*). In cotton (*Gossypium hirsutum*), the fibre is the hairs that are attached to the seeds.

**Figure 3.11**

Ropes have traditionally been made from the natural fibres of a range of plants, such as coir from the coconut palm *Cocos nucifera*, and sisal from *Agave sisalana*, shown here being harvested in Madagascar.

For thousands of years, a variety of plants including cotton, flax and hemp have been grown to provide fibre, mainly for clothes and ropes. Although clothing fibres are very important, this section discusses other, possibly less familiar, uses of fibres. The use of plant fibres to make ropes was a major industry in the days of sailing ships, which required large amounts of rope for the rigging of sails.

**Figure 3.12**
Some plant fibres, such as those of (a, c) flax, are soft and possess a fine sheen whilst others, like those of (b, d) hemp, are much harder and coarser.

All plant fibres are flexible, have very little elasticity, have good resistance to damage by abrasion and can withstand both heat and sunlight. Coir fibre from the coconut fruit resists sea water damage, but like all natural fibres, it rots if soaked over a long period because of the action of micro-organisms. Currently, however, the majority of worldwide industrial rope-making uses oil-based synthetic fibres such as nylon and polypropylene. These are stronger than the strongest natural fibre, hemp, and are not damaged by micro-organisms but have other disadvantages: they deteriorate in bright sunlight, melt at a lower temperature than natural fibres and are not biodegradable, so their disposal can be an issue.

The advent of human-made fibres in the latter half of the twentieth century reduced the demand for plant-based fibres. However, growers and exporters of plant fibres have sought to develop new markets for their products and the demand for plant-based fibres for more diverse commercial applications is beginning to increase. One such example is the use of leaves of abaca (*Musa textilis*), a close relative of the banana, which is a native

a

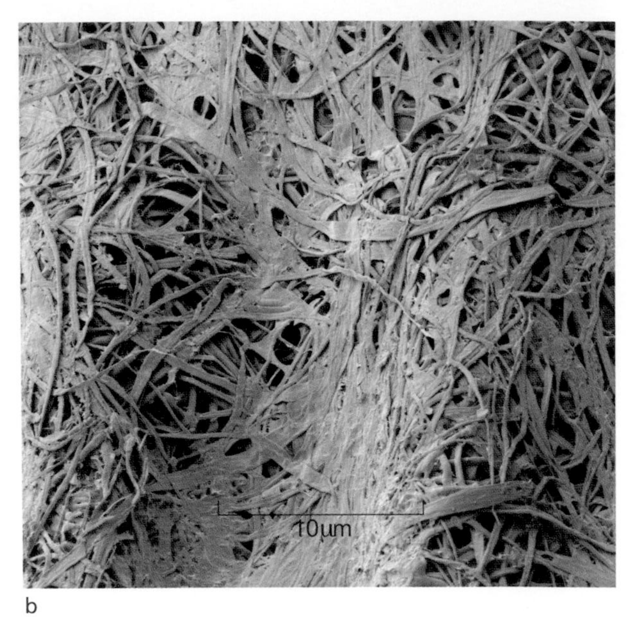


b

**Figure 3.13**

(a) Abaca (*Musa textilis*) provides extremely strong and flexible fibres that have many uses.

(b) In this coloured scanning electron microscope image you can see the fibres in tea bag paper, which is made from a blend of wood and abaca fibres.

plant of Borneo, long grown in the Philippines and also grown on plantations in Ecuador. Commercially known as manilla hemp, the abaca plant matures in 18–24 months and its leaves can be harvested three times a year. Both the Philippines and Ecuador export abaca pulp and fibres across the world, and this is a key component of the increasing economic prosperity of these countries. Abaca leaf fibre contains up to 15% lignin, making it extremely strong and waterproof. It has traditionally been used to create ropes for house building; this use is ongoing but abaca fibre is now also valued for fishing lines, sacking, textiles and for generating a fine, flexible paper that can be used for banknotes, tea bags, the original manilla envelopes and cigarette papers. The fibres have also been shown to have great potential in the production of car body parts.

Most recently, Mercedes Benz, alongside a number of other car manufacturers, have taken abaca fibres and mixed them together with a special polypropylene thermoplastic that can be moulded when heated to create body parts for some of their luxury cars. It has been estimated that this uses 60% less energy than the production of the traditional material fibreglass.

**Figure 3.14**

Banknotes in both Japan and the Philippines contain a high percentage of abaca fibre.

## The world's first Formula 3 green racing car

The first Formula 3 green racing car was launched in 2009 as a result of research carried out by the Warwick Innovative Manufacturing Research Centre (WIMRC), which is part of Warwick University, UK. The motor-racing industry is often thought of as excessive in terms of resources used. The WorldFirst racing car is the first to be created using sustainable and renewable materials. These include a steering wheel created out of a polymer (a large chemical molecule made up of a series of repeating smaller units) made from carrot and other root vegetable fibres, and flax fibres in wing mirrors, wings and the body shell. Other innovative materials include potato starch, used as a filler inside the body panels, and soybean foam. The biodiesel fuel and lubricants are also derived from a variety of plant sources, including chocolate. The £300,000 car is not just for show; it is functional and can reach speeds of over 200 kilometres per hour. The creators hope that the car will be developed further, for instance, the safety critical parts of the car are not made from sustainable materials. Perhaps the car, or the technology developed from it, will be used in the future by a major Formula 1 team.

**Figure 3.15**
The WorldFirst Formula 3 green racing car is a demonstration of the potential use of plants in the automotive industry.

# 3.4 Starch

Any photosynthetic sugars that the plant does not use quickly are stored, mainly as starch. Starch is familiar to us as the carbohydrate found in grains, legumes, potatoes and bananas. Starch provides a source of stored energy in seeds, which can be utilised after germination to sustain the young plant until its photosynthetic ability reaches a level whereby it becomes self-sufficient.

## What is starch?

Glucose is a sugar that is made through the process of photosynthesis and starch is a polymer composed of chains of individual glucose molecules, called glucose monomers. There are two forms of starch, amylose and amylopectin. Amylose is a simple, linear chain of glucose molecules linked one to another. Amylopectin also comprises glucose molecules linked in a linear fashion, but additional glucose molecules form branches to the chain. Amylopectin molecules are larger than amylose molecules and can be made up of hundreds of thousands of glucose molecules.

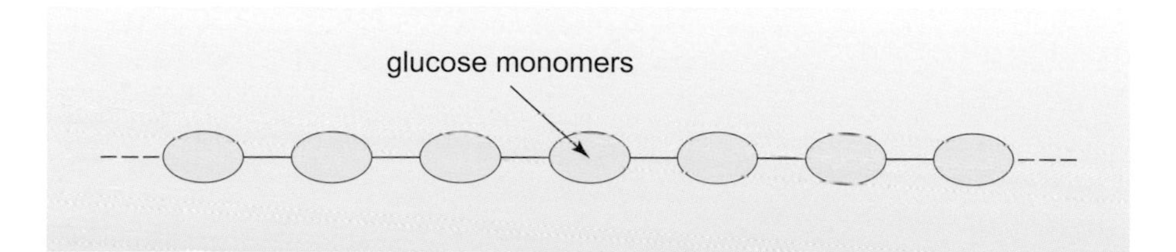


**Figure 3.16**
The structure of a starch polymer showing how glucose monomers are joined to one another in amylose.

Although some starch can be formed in the chloroplasts, where photosynthesis occurs, most starch production occurs in the storage tissues of plants, such as tubers or seeds. It is not a coincidence that these tissues are the very parts of the plants that humans harvest. Many of the world's staple foods are rich in starch, such as rice, wheat, maize, cassava, yam, sorghum and potatoes.

**Figure 3.17**
The most important staple crops in the world all contain large amounts of starch.

cassava

rice

sorghum

yam

## The uses of starch

Starch has been used for both food and non-food products for centuries. Clearly, its major current use is as a foodstuff. As well as being the key component of grains, much of the world's production of starch is converted to syrups for use in the food industry. Starch syrup is a major ingredient in a large number of different food products, for example in pre-prepared baby foods.

### High fructose corn syrup (HFCS) and possible health concerns

HFCS originates from maize (*Zea mays*), often called corn in the USA, and is produced by converting the starch present in the cob into sugar, first glucose and then into fructose, a very sweet sugar found in honey and ripe fruits. Typically, this fructose-enhanced product is mixed with pure corn syrup, containing 100% glucose, to produce a product (HFCS) that has the correct degree of sweetness. HFCS is both easy to transport, as it is a liquid, and also cheap. In the USA, HFCS is used in a wide range of processed food and soft drinks in greater amounts than sugar derived from sugar cane or beet. The increasing use of HFCS has been linked to rising rates of obesity in the western world. Concern has also been expressed over the labelling of foods containing HFCS, as it is not necessarily obvious to consumers that in reality HFCS means sugar.

Bioplastics are biodegradable and compostable alternatives to petroleum-based plastics and are made from raw plant materials, particularly starch. The most commonly used is corn (maize) starch, which can be broken down by bacteria to produce lactic acid. This can then be polymerised using a catalyst to produce a colourless product known as polylactic acid (PLA) plastic. This degradable plastic has been used for several years in medicine as dissolvable sutures but is also gaining commercial interest because of its potential use in products such as shopping bags, food packaging and microwavable food trays.

**Figure 3.18**

Increasingly, starch-based alternatives are being used as a substitute for synthetic materials such as plastic and polystyrene. Starch derivatives can be formed into cups and containers and are increasingly used in product packaging, as shown here.

The increasing interest in bioplastics is driven by rising oil prices (making oil-derived plastics more expensive) and by concerns for the environment. However, there is significant debate as to how disposable these bioplastics are. One view is that they can only be fully broken down when special digesters that consume large amounts of energy are used. As part of this digestion process, methane gas is released and this is one of the greenhouse gases linked with climate change. Another challenge to be faced when manufacturing bioplastics is that the cultivation of crops for the plastics may reduce the land available for food production.

**Figure 3.19**

Maize (*Zea mays*) cobs come in a variety of shapes and colours. The starch which can be extracted from maize is the main focus for the production of bioplastics.

# 4 Biofuels

The use of biological material (biomass) as a fuel or lighting material is not new; the earliest evidence for humans using wood for fires is at Gesher Benot Ya'aqov in Israel, 790,000 years ago. However, concern over the impact that the burning of fossil fuels is potentially having on the climate has resulted in renewed interest in biofuels. The difference between fossil fuels and biofuels is that fossil fuels were produced millions of years ago when plants and other organisms died, became buried and were subjected to high temperatures and pressures forming coal, oil or natural gas. Biofuels, on the other hand, are produced from biological material that has been living recently. There are a number of ways in which biofuels can be produced.

Some biofuels can be produced from waste material, such as recycled plant oils, whilst others can be produced from plants specially grown for the purpose. Both liquid and gaseous forms of biofuels can be produced from crops that either have a high sugar content, such as sugar cane or sugar beet, or contain starch that can be converted into sugars, such as maize. Plants containing high levels of plant oils, such as oil palm or soybean, can also be used. Wood and its by-products can be converted into a variety of biofuels.

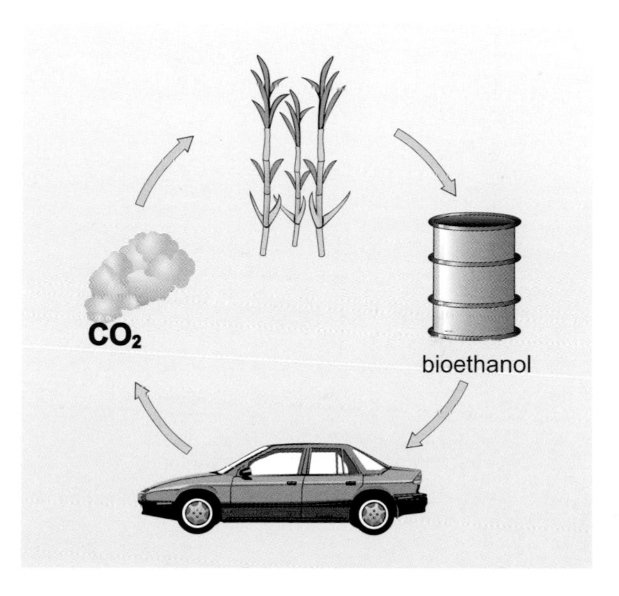


**Figure 4.1**
General principle of biofuel production.

## 4.1 Energy from plants and climate change

Plants capture carbon dioxide ($CO_2$) during photosynthesis and store it in their tissues in a variety of carbon compounds (including both carbohydrates and oils). These are often energy stores for the plant but they can also be harvested and processed to provide energy for human use.

### Comparison of the energy content of plant products

| energy store | energy (megajoules per kilogram) |
|---|---|
| plant oil | 37 |
| coal | 24 |
| carbohydrates (including sugars) | 17 |
| wood (dry) | 16 |

1 megajoule is roughly the amount of energy a one-bar electric fire would emit in 15 minutes. (Note that if wood has not been seasoned to remove excess water less energy is obtained, as some of the energy is used to heat and evaporate the water.)

## The carbon cycle

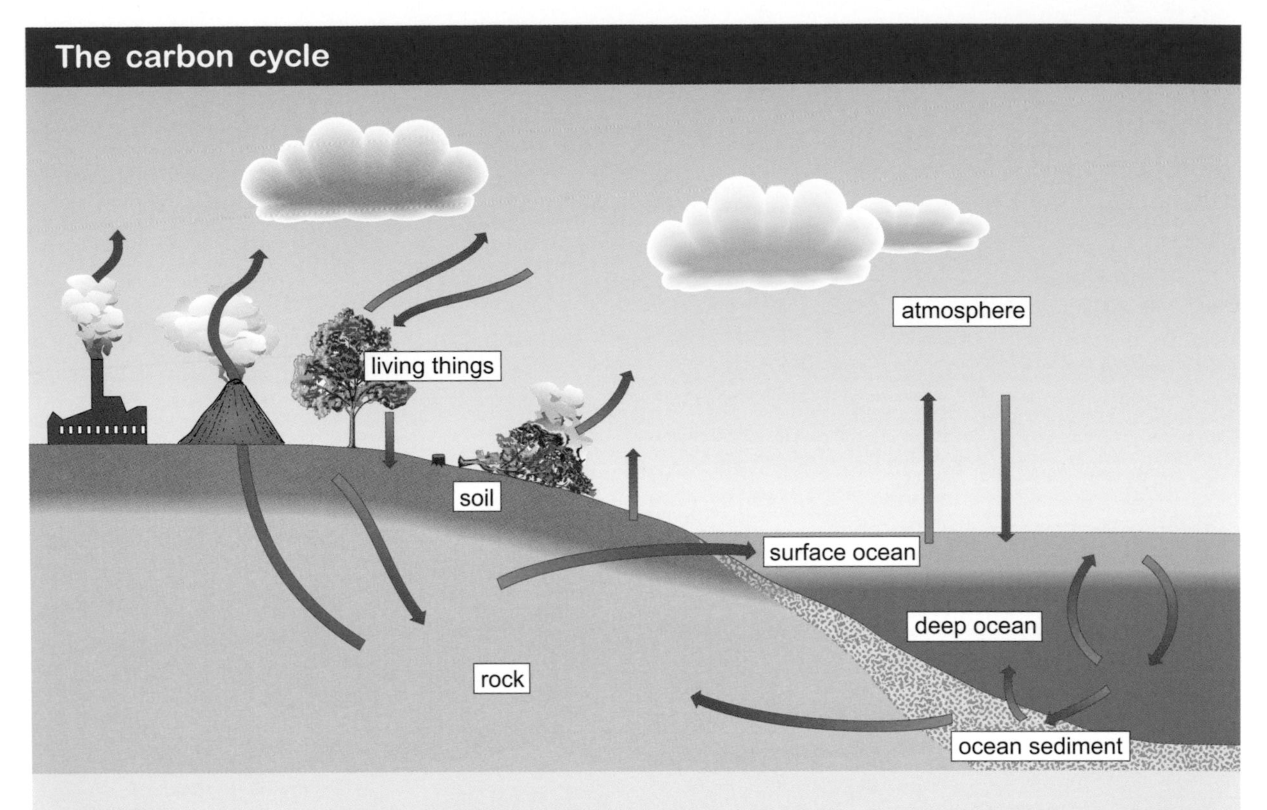


**Figure 4.2**
Simple representation of the carbon cycle.

All living organisms, not just plants, have structural components that are based on carbon. The pictorial presentation of a carbon cycle illustrates how carbon moves between the reserves found in living things (plants and animals), soil (decomposing organic matter), rocks (including fossil fuels), the atmosphere and the oceans.

At present, most of our energy comes from fossil fuels which originate from $CO_2$ 'locked up' in plants by photosynthesis millions of years ago. When fossil fuels are burnt, $CO_2$ is released into the atmosphere adding to the levels already present. Since the industrial revolution, there has been an acceleration in the burning of fossil fuels. The levels of $CO_2$ in the atmosphere are monitored by the Mauna Loa Observatory in Hawaii and are estimated to have risen from 280 parts per million (p.p.m.) in 1800 to 387 p.p.m. today (i.e. in 2009). The increase in $CO_2$ in the atmosphere has been linked to global warming (the increase, of around 0.5 °C, in the average temperature of the Earth since around 1920). $CO_2$ is known as a 'greenhouse' gas and it acts with other greenhouses gases in the atmosphere (water vapour, ozone, nitrous oxide and methane) to insulate the Earth. Increases in these greenhouse gases increase this insulation and result in rises in the Earth's temperature. The predicted dire consequences include extreme weather conditions and melting of ice sheets and glaciers, resulting in rising sea levels and hence coastal flooding.

The Kyoto Protocol from the United Nations Framework Convention on Climate Change, which came into force in 2005, legally commits countries that signed the protocol to reduce emissions of four greenhouse gases, including $CO_2$. One of the ways in which this could occur is by a shift away from fossil fuels to biofuels, although this view is contentious for reasons indicated shortly. Another is by the process known as carbon offsetting, whereby the amount of carbon released into the atmosphere by burning fossil fuels is balanced, for example by planting a certain number of trees to take up the equivalent amount of $CO_2$ by photosynthesis – so-called carbon sequestration. This has led to the currently used terms 'carbon neutral' (where the amount of $CO_2$ that is produced by an individual or a population is balanced by the amount of $CO_2$ that can be absorbed through various measures taken) and 'achieving a zero carbon footprint' (where the net amount of $CO_2$ released by a person, or by an object such a house, is zero).

## 4.2 Wood as an energy source

The use of wood as fuel for cooking and heating has a long history. In developing countries, 50–90% of fuel used for such purposes comes from either wood or other plant biomass. However, wood is increasingly providing a fuel for electricity generation. The burning of wood is considered to be carbon neutral as it does not release more $CO_2$ into the atmosphere than if the wood were to decompose naturally, although this $CO_2$ is released in a very short space of time.

Woodlands can be managed sustainably to allow appropriate harvesting and replanting; this also provides some local employment and creates a pleasant and enjoyable place to visit. Woodlands also perform a useful function by temporarily storing rainwater, thereby preventing excessive water from entering streams and rivers too quickly. Wood burning also provides an outlet for wood residues from forestry activities which would otherwise end up in landfill sites. There is concern, however, that there will not be enough material to sustain increasing demand for woodfuels.

The process of producing heat and electricity from woodfuel is complex. First, the wood has to be dried, then it undergoes pyrolysis (heating in the absence of oxygen) to produce gases. The gases are then purified and burnt to generate electricity. The ash created during pyrolysis contains nutrients which potentially could provide a plant fertiliser, though it is possible the ash may contain contaminants from the soils in which the trees had originally been growing.

Potential sources of woodfuel are many and include early thinnings from commercial plantations, the residues from timber harvesting and arboricultural activities, coppicing and sawmills. In addition, wood pellets made from highly compressed waste sawdust are gaining in popularity as domestic fuels in the USA and various Scandinavian countries.

Some wood harvesting systems use the whole tree for chipping, while others utilise only the stem wood. In 'short rotation forestry', fast-growing trees are cultivated and grown until they reach what is considered an economically optimum size; the time this takes varies depending on the tree species. The trees are then either harvested or coppiced. Harvesting involves completely cutting down the trees, possibly removing the roots, and then replanting with saplings. In coppicing, the young stems are cut back to encourage a number of new stems to grow from the 'coppice stool'. This can help increase levels of carbon dioxide uptake (carbon sequestration) as the coppice stool re-grows.

## Short rotation coppice: willow and poplar

Willow (*Salix* species) and poplar (*Populus* species) are fast-growing trees, and because they can be densely planted, they give a high yield of wood in a relatively small area. Stems can be coppiced every 3–5 years and the coppice stools remain productive (produce new stems) for up to 30 years before the root stocks need to be replenished. There is much research interest in short rotation coppicing and the collaborative 'Biomass for Energy Genetic Improvement Network (BEGIN)' have undertaken a number of trials to find varieties and cultivars of willow and poplar species that offer high stem yields, increased growth rates, increased pest and disease resistance and high energy output when burnt.

Establishing willow or poplar plantations has other benefits, such as creating a method of diversification for agricultural land use, increasing biodiversity and providing shelter or screens both for wildlife and against pollution. Willow, in particular, is very good at taking pollutants, including excess nitrate from fertilisers, out of the soil (a process known as phyto-remediation) but this can lead to a high risk of contaminants in the wood.

a

b

**Figure 4.3**
(a) A farmer inspects his plantation of fast-growing willow (*Salix*).
(b) A newly coppiced oriental plane tree (*Platanus orientalis*).

## 4.3 Grasses as an energy source

There is increasing interest in using grasses as a biofuel source. Grasses grow quickly, produce a large amount of biomass per unit growing area and leave only small amounts of residue when they are burned. Typically, grasses are burned to produce heat and steam to power turbines in conventional power plants. The largest power station in the UK, Drax, currently burns 300,000 tonnes of *Miscanthus* x *giganteus* annually, alongside its usual coal supplies. Another advantage of using grasses is that many of them multiply by underground rhizomes, which means they spread quickly and generate new shoots easily. However, this can also make them invasive and difficult to control and eradicate, so they are considered by some to be weeds.

**Figure 4.4**

Canary grass (*Phalaris canariensis*) originated in the Mediterranean region but has been cultivated and grown in the Middle East, Europe and parts of Argentina. It was formerly cultivated primarily for the production of bird seed but is now being considered as a biofuel, due to its high yields and ability to thrive in temperate regions. Breeding programmes have selected varieties which have a particularly high carbon content in their cell walls, giving a potentially greater energy output when burned.

**Figure 4.5 - far left**

*Miscanthus* x *giganteus* is a large perennial grass that can grow to a height of over four metres. It is grown in the UK and elsewhere in Europe and for several years there has been intense research into its potential as a biomass crop. It reproduces by underground stems (rhizomes) and is considered to be an environmentally friendly crop due to its large root system, which captures nutrients, and its stems, which provide wildlife cover.

**Figure 4.6**

Switch grass (*Panicum virgatum*) is a dominant species of the tallgrass prairies in North America and is commonly used as an ornamental grass in gardens. It thrives in harsh conditions, needing only poor soil with a low nutrient content, and undergoes rapid growth. These characteristics make it a strong candidate for biomass production.

There has been interest in using mixtures of a number of plant species, including grasses, together to produce biofuel. These mixtures allow the maximum interception of light as the different plants grow at different rates. They are considered to be carbon neutral as they are perennials, which do not require re-sowing each year, and, most importantly, they can be sown on degraded agricultural land so food production is not jeopardised.

## 4.4 Transport biofuels: biodiesel and bioethanol

Biodiesel is an alternative fuel to diesel, and bioethanol can be used in place of petrol. Biodiesel cannot be used on its own in traditional internal combustion engines but can be blended with traditional petroleum fuel. Bioethanol can be used on its own or in a blend. Overall, biofuels make up 5% of the petroleum blend in the UK. The UK Government set a target that 5% of road transport fuel should be from renewable (non-fossil fuel) sources by 2010.

**Figure 4.7**

Due to increasing demand for transport fuel, alternatives to petrol and diesel are under scrutiny.

Biodiesel is produced from oil-seed crops, such as oilseed rape, sunflower oil, palm oil and soybean oil. Oilseed rape is the main oil crop in the UK and yields approximately 1,300 litres of biodiesel per hectare planted. The fuel is synthesised from the oil seeds by mixing them with two industrially sourced chemicals, methanol and either sodium hydroxide or potassium hydroxide. This reaction causes the formation of fatty acids and glycerol, with the fatty acids reacting further to produce biodiesel. Most biodiesel production is currently in Europe, but it is not limited to the developed world. Tanzania, for example, has started producing biodiesel from the nuts of the croton tree (*Croton megalocarpus*), a plant native to South-East Asia. Several countries, notably India, are using biodiesel obtained from *Jatropha curcas* oil as a fuel; indeed the trains that run between Mumbai and Delhi now use diesel that contains 15% biodiesel obtained from *Jatropha*. This plant has several advantages as a source of biofuel: it can tolerate drought conditions, and it grows on poor land that is unsuitable for agriculture and so does not compete with food production.

a

b

**Figure 4.8**

(a) Sunflowers are just one oil crop that is being investigated as a potential biofuel.

(b) Tropical plant species, such as *Jatropha curcas*, could also be a potential source.

Germany used significant quantities of bioethanol as a biofuel during the Second World War, but recently interest in this fuel source has spread. Currently, countries using bioethanol include Brazil, France, the USA, the UK, Argentina and South Africa. In Brazil use has been particularly widespread; however, this has been at the expense of large tracts of natural vegetation including rainforests. It is estimated that one-third of cars in Brazil run entirely on bioethanol and the remaining two-thirds use a mixture of bioethanol and petrol (gasoline). Bioethanol is produced by fermenting plant sugars, or it can be made by taking starch-rich plant material, treating it to convert the starch to sugars, and then fermenting them. A third, less productive, method is to take stems or leaves and convert the tough cellulose from cell walls into sugars for fermentation; but this requires additional steps, including treating the plant material with sulphuric acid and heating to convert the cellulose into sugars, or using enzymes to do this. Both sulphuric acid and the necessary enzymes have to be produced by industrial means and this incurs an extra cost.

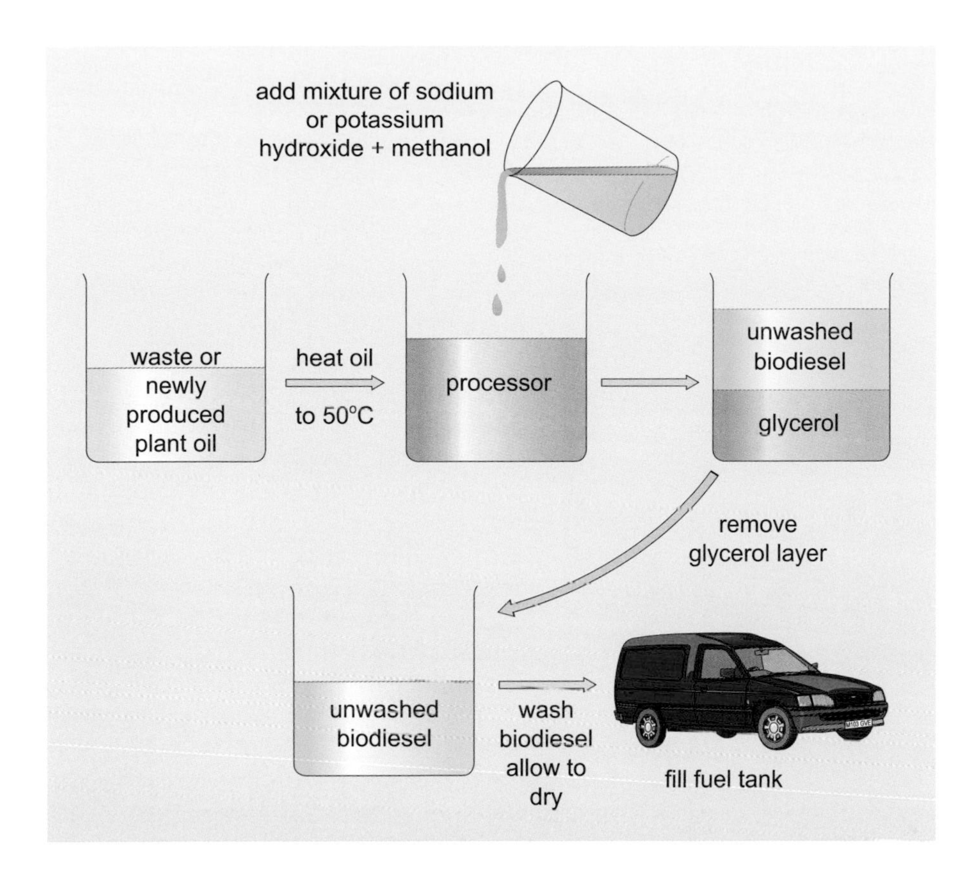


**Figure 4.9**

The production of biodiesel from waste cooking oil or newly produced plant oil is increasingly gaining favour in the developed world. However, the amounts produced are currently very small compared with conventional petroleum.

**Figure 4.10**

The steps involved in the production of bioethanol.

## 4.5 Biogas

Biogas is mainly a mixture of methane and carbon dioxide produced through the breakdown of organic materials by micro-organisms in the absence of oxygen (that is, under anaerobic conditions). Organic materials can include municipal waste, food and animal waste, sewage, and biomass crops such as switch grass and *Miscanthus*. Biogas produced in sewage treatment works can be used to generate the electricity to power the works.

**Figure 4.11**

A public bus running on biogas, in Vasteras city centre, Sweden.

Production of biogas from food waste reduces the amount of this waste that reaches increasingly scarce landfill sites, as once the gas has been produced any remaining residue can be used as agricultural fertiliser. Biogas also enables a country to produce its own fuel, rather than being dependent on foreign imports. Sweden is currently the world leader in the use of biogas, having the world's first biogas-powered train which runs the 100 kilometre route from Linkoeping to Vaestervik. Although it is powered by biogas produced from animal waste, there is no reason why in the future such transport could not be powered by biogas produced from plant sources. Buses and rubbish trucks in several Swedish towns also run on biogas.

## 4.6 Biofuels: some of the issues

### How does the cost of biofuels compare with that of conventional fuels?

At present, the cost of producing biodiesel and bioethanol is approximately double that for conventional fossil fuels. The UK offers tax incentives to encourage fuel producers to make biofuel and there is also an incentive for consumers: the duty on petrol with 5% biofuel is, at the time of writing (2009), 20 p/litre less than that on ultra-low sulphur petrol.

A recent review has shown that biofuel production is a high-cost process overall; only the production of bioethanol from sugar cane was considered to be cost-effective. If, however, the price of crude oil rises then biofuels may start to become more competitive, although this would depend on fertiliser, and hence biofuel feedstock, prices remaining stable.

### Does growing crops for fuel create higher food prices?

One point of view is that using land to grow crops for fuel rather than for food has led to increased food prices, which in turn impacts most on those in developing countries who have to spend a relatively large proportion of their income on food. In 2007, the United Nations Food and Agriculture Organisation (FAO) and the International Food Policy Research Institute (IFPRI) both released reports suggesting that rising food prices were due to the conversion of agricultural land from food to biofuel crops. Subsidies for the biofuel industry, quoted to be in the region of £5.6–6.1 billion per year (FAO), are believed to incentivise farmers to grow biofuel rather than food crops. However, those supporting biofuels hold the view that better production methods, which will use parts of the plant currently wasted in biofuel production, may be able to increase the efficiency of biofuel production.

## Does growing biofuel crops result in more fertiliser and pesticide input?

The amount of fertiliser and pesticide input depends on the crop species grown; one criterion for a suitable biomass crop is that it has a low requirement for nutrients. Switch grass, for example, grows well in poor soil conditions and requires around two-thirds less fertiliser input than a food crop such as maize.

## Is there a potential loss of biodiversity?

In some cases, the use of biofuels could increase biodiversity (the number of different species found within a given area), for example, by harvesting woodfuel and then replanting with mixed woodlands. Conversely, the increased use of soybeans for biofuel in the USA has caused soybean production to be increased in South America at the expense of Amazonian rainforest, which has been cut down to give more agricultural land. This has the potential to reduce biodiversity. Also, trees in rainforests help absorb $CO_2$ (albeit not at the rate of newly planted forests) and so are believed to play a part in offsetting global warming.

## Is the energy output greater than the energy input?

The overall energy balance, that is the net energy output compared to the input needed, is dependent on both the species of biomass crop and the type of biofuel being produced. The inputs to a biofuel crop include both the direct costs of growing, such as fertiliser, cultivation, harvest and labour, and the indirect costs, such as storage of the harvested crops. Processing inputs include pre-treatments, processing for energy release and the removal or reuse of residual waste. Ideal crops are being investigated that require low maintenance, need only low nutrients, grow quickly and can tolerate poor soil conditions. Switch grass is one such plant, and interestingly it has been shown that it is only energy efficient if it is used as a pelleted biomass for fuel rather than for bioethanol production.

## Do biofuels reduce carbon emissions?

Burning biofuels does release $CO_2$, but this is $CO_2$ that has been removed from the atmosphere relatively recently, unlike the carbon in fossil fuels which has been removed from the atmosphere for many millions of years. Hence, biomass energy crops are deemed carbon neutral. However, some argue that nitrous oxide, which is 240 times more powerful as a greenhouse gas than $CO_2$, is produced in greater quantities during the production of biogas than during the burning of fossil fuels. Also, biofuel production generates $CO_2$ through the use of agricultural and haulage vehicles for growing and transporting the crop.

# 5 Plants in crime

contributed by Patricia E. J. Wiltshire

Toxic substances from plants have been used by criminals for hundreds if not thousands of years, but recently the study of plant-related material at crime scenes has allowed plants to be used to solve crimes. Scientists who study such crime scenes are forensic botanists, and they draw on knowledge of the anatomy, taxonomy, ecology, and distributions of plants, as well as of pollen grains and spores. (Spores are produced by the most simple non-flowering plants, such as mosses, liverworts and ferns, to allow them to reproduce.)

Plants, or parts of plants, have been used with criminal intent against people, but they have also been used to:

- detect the routes taken by the perpetrator at crime scenes
- estimate the time a dead body was placed on or in the ground
- provide information on the gut contents of a corpse
- find the bodies of people who have been murdered and subsequently buried or left on the surface of the ground
- show links between objects, places and people
- detect drug abuse.

## 5.1 Plants as poisons

Plants produce a bewildering array of compounds, some of which are poisonous to humans and other animals. Certain plant toxins are heat-sensitive and will be rendered ineffective by cooking or other processing, but others retain their toxicity. Some animals are capable of detoxifying certain compounds. For example, although the red fleshy covering surrounding the berry of European yew (*Taxus baccata*) is not toxic to humans, the black seed in the centre is highly poisonous. Yet many birds eat the whole fruit with no apparent harm and field mice seem to gorge themselves on yew berries without apparent ill-effect.

It must be stressed that the term poisonous is a relative one, and the toxicity of any plant species depends on the part of the plant in question, the levels of toxin within the plant tissue, and the susceptibility of the victim. Toxins are often contained within only part of a plant, so the active substance may be confined to leaves, flowers, stems, roots, or even bark. Even edible plants may have toxic tissues. For example, the leaf stalks of rhubarb (*Rheum rhaponticum*) are cooked and eaten, but the leaves themselves have high levels of oxalic acid which is, amongst other things, a nerve toxin. In some cases, hairs on the stem or leaves, or even secreted fluids (known as exudates), can be toxic; in others, compounds may become poisonous only under certain circumstances. In humans, the response to such substances can be highly variable – a low concentration of a 'toxin' may be beneficial to human health whereas a higher concentration of the same substance can be lethal. It is important to distinguish between toxicity and lethality.

**Figure 5.1**
All parts of oleander (*Nerium oleander*) are poisonous whether cooked, fresh or dried.

## Oleander

Oleander (*Nerium oleander*) is a native shrub of southern Europe but it can survive without winter protection in the milder parts of Britain. Oleander contains a number of chemical compounds called glycosides which have effects similar to those of digoxin, a compound found in the foxglove (*Digitalis purpurea*) that is used as a medicine to slow down the heart rate. In humans, most cases of reported deaths from oleander have been through accidental poisoning of children, or where adults have mistakenly used the leaves as herbal medicines. Intentional poisoning of adults has also occurred, with as few as seven leaves causing death.

## Monkshood (wolf's bane)

Monkshood (*Aconitum napellus*) is a herbaceous perennial grown for its spikes of blue hooded flowers. The plant is toxic when eaten but the active agents can also be absorbed through skin. It is one of the most poisonous species in Britain and is known as 'the queen of poisons'. The plant contains a number of poisonous chemical compounds, including one called aconitine. This is thought to be a particularly strong poison but other toxins have also been identified in the plant. They cause a number of symptoms including: burning of the mouth, abdominal pain, delirium, reduced heart rate, paralysis, coma and respiratory failure.

**Figure 5.2**
Monkshood (*Aconitum napellus*).

In 2002, a case was reported of a 60-year-old male who committed suicide by swallowing a preparation of monkshood. After taking the poison, he became agitated, nauseous and vomited violently. He told doctors what he had done but they were unable to save him and he died of cardio-respiratory arrest two hours after he had swallowed the monkshood.

At the Old Bailey in London in February 2010, an Indian woman was convicted of poisoning her secret lover to prevent him marrying another woman. She gained access to the man's home and put some monkshood in a curry she found in the refrigerator. Within hours of eating the food, her ex-lover was dead and his fiancée was critically ill. The defendant was given a long custodial sentence for murder.

a

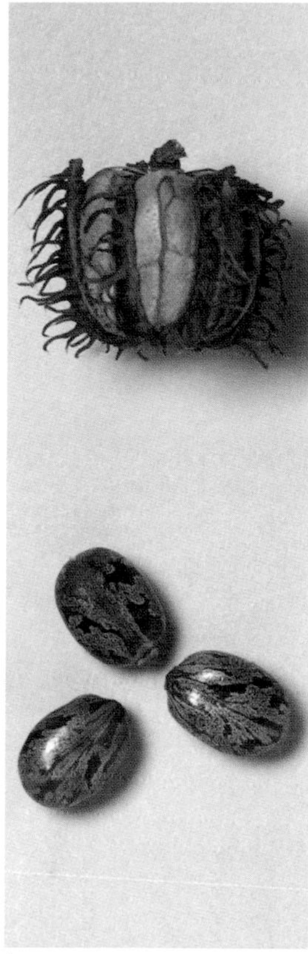

b

**Figure 5.3**
Ricin, just one of several poisonous compounds produced by (a) the castor oil plant (*Ricinus communis*), is about 12,000 times more poisonous than rattlesnake venom. (b) The seedpod and seeds of this plant.

## Castor oil plant

The castor oil plant (*Ricinus communis*) is an annual herbaceous shrub, grown commercially for its seeds, which yield castor oil, and as a decorative garden plant. The plant produces at least six compounds that are poisonous to humans. The most toxic are ricin, ricine and ricinine. Ricin is a complex compound which prevents the manufacture of proteins. Ricine coagulates blood cells, while ricinine damages the nervous system.

There is wide variation in sensitivity to ricin in different animal species but, in all cases, the lethal doses are very small. A dose as small as two-millionths of the body weight would be fatal by injection, although the lethal oral dose would be greater.

### Killed by an umbrella?

Ricin was used to murder a Bulgarian dissident, Georgi Markov, in 1978. He was a BBC World Service journalist and a strong critic of the communist regime. While he was at a bus stop on Waterloo Bridge in London, he felt a sting on the calf of his right leg. By evening, he had developed a high fever and he died three days later. Because the symptoms of ricin poisoning are similar to those of septicaemia (blood poisoning), he was diagnosed incorrectly. When police were told that he had been receiving death threats, they investigated further and a small platinum pellet was found in his calf. The pellet had been drilled to contain the poison, which had seeped into his bloodstream after he had been injected. Some accounts suggested that his leg had been injected by the tip of an umbrella.

## 5.2 Use of plants in forensic science

The walls of plant cells are composed of cellulose; in xylem cells, the walls are also impregnated with lignin. This gives the cells great strength and makes them resistant to decomposition. In pollen grains and spores, the cellulose walls are protected by an outer wall, called the sporopollenin. If conditions are unfavourable for microbial activity (as, for example, in very dry, acidic conditions, or conditions of low or no oxygen), cellulose, lignin and sporopollenin can persist for very long periods, even many thousands of years. This means that they can remain identifiable long after crimes have been committed and provide evidence for criminal investigators even when offences happened many years previously.

A study of the physical characteristics of wood can enable links to be made between fragments of wood and original items. A classic case was that of the abduction and death of Charles Lindbergh's son in 1932 in the USA. The prime suspect for the crime was Bruno Richard Hauptman. The offender had used a home-made wooden ladder to gain access to the child's nursery, and a botanist identified the four kinds of wood used in the ladder. These were

### Trace evidence

**Figure 5.4**

Plant habitats and communities do not only occur in nature: every garden forms a plant community, and in some crime cases these unique plant combinations can provide valuable forensic evidence.

Plants also provide very good 'trace evidence'. In 1910, Edmond Locard established the principle that 'every contact leaves a trace'. In nature, organisms perform best in their preferred habitats, and those with similar requirements will be found growing together in recognisable and predictable communities. The habitat and the community together make up the ecosystem. Natural and semi-natural plant ecosystems have a degree of predictability, but none is identical. Every location has a peculiar and distinct 'biological signature'. Gardeners create habitats and communities, although these depend on the whims of the individual rather than any natural distribution. But the very artificiality of the garden ecosystem can be valuable in the forensic context, since no two gardens contain the same plant combinations.

Any ecosystem can be recognised from little fragments of it. These can be pollen grains, plant spores, fragments of whole plants (leaves, flowers and stems), egg shells, snail shells, soils, fragments of insects, and so on. Fragments such as these are called 'proxy indicators' and they are useful for linking any object to locations that might be important in a criminal case. However, interpretation of proxy evidence requires considerable knowledge on the part of the analyst.

then matched with shipment records to specific suppliers and to wood from Hauptman's attic. Furthermore, tools belonging to Hauptman were found to have made distinctive marks on the ladder. The wood analysis provided compelling evidence for the prosecution.

Animals and plants have predictable life cycles and patterns of growth, and these can be exploited to estimate the timing of events. The counting of growth rings in woody stems and roots that have grown through buried remains is an obvious indicator of time. However, in cases of human murder and subsequent burial where the body is in a very shallow grave, roots are frequently too young to be useful.

Plants respond to disturbance such as trampling or digging. Different plants often invade disturbed areas so that an affected area can have a different plant community from its surroundings. However, these changes are often very subtle, and finding clandestine graves is not an easy matter. Plants may also respond to damage by undergoing corrective growth with prolific side-shoot development, or they may have their growth arrested during the growing season by seepage of body fluids. The thickness and nature of leaf litter below and over a corpse and the relative growth of perennial plants in relation to the remains are also useful. All these phenomena can also help in estimating the length of time a body has lain in a place.

### The Soham murders

In 2002, the bodies of two schoolgirls from Soham, Cambridgeshire were found in a drainage ditch running alongside a grassy track, close to Lakenheath Airbase in Suffolk, UK. The author[1] was asked to find where the murderer had entered the ditch. This was not obvious since the banks along the grassy track supported dense growths of nettles (*Urtica dioica*), grasses (Poaceae), hedge woundwort (*Stachys sylvatica*) and hogweed (*Heracleum sphondylium*). By careful examination of individual plants along the edge of the track, a faint path was identified. This was marked by corrective growth of various plant species. From estimating the growth rates of these plants' side-shoots, it was possible to estimate the length of time between deposition of the bodies and their discovery.

The main suspect was Ian Huntley, who worked as a caretaker at the victims' school. He had hidden the girls' clothing in a dustbin in one of the school buildings and these contained leaf fragments, pieces of twig and fruits of plants growing in and around the ditch. Pollen from the scene was also found on two pairs of Huntley's shoes, in his car and on plastic petrol cans. There was evidence that he had visited the site twice before the girls were found.

[1] Patricia E. J. Wiltshire

## 5.3 Forensic palynology

The forensic palynologist studies pollen and spores, and any other microscopic entity that may find its way into a palynological sample. Although not derived from plants, fungal spores have enhanced forensic palynology greatly in recent years, and many have distinctive structures.

Pollen and spores can be transferred directly from plants or fungi, from soil and vegetation, or from any surface which has accumulated them. Under certain circumstances, they can also be secondarily transferred to other objects.

## Pollen structure

Some pollen grains, such as those from pine species, have air sacs to aid dispersal through the air, or other outgrowths, as in another conifer *Cryptomeria*. Most pollen is dispersed as single grains, although the pollen of some plants, such as the heathers (Ericacae) and the genus *Mimosa* (legume family), is dispersed in groups of four or even more.

Most pollen has an outer covering which is perforated by pores or furrows, and these can be present in a variety of sizes and combinations. Pollen can also be variably patterned, and this may relate to its mode of dispersal. For example, in some wind-pollinated plants, such as grasses (Poaceae) and hazel (*Corylus*), the outer covering is smooth; insect-pollinated grains can be highly sculptured which may help the pollen to cling to its animal pollinator. The sculpturing can be simple or complex and can take many forms. The elements making up the patterning can be in various combinations, even in one pollen type. Some have a network of ridges (reticulum) as in willow (*Salix*) and geranium. Others have spines of various sizes on part of, or all over, the grain surface as in thistles (*Cirsium*); while dandelions (*Taraxacum*) have coarse spines arranged on top of an elaborate,

**Images in Figures 5.5 – 5.8 were all taken with a scanning electron microscope.**

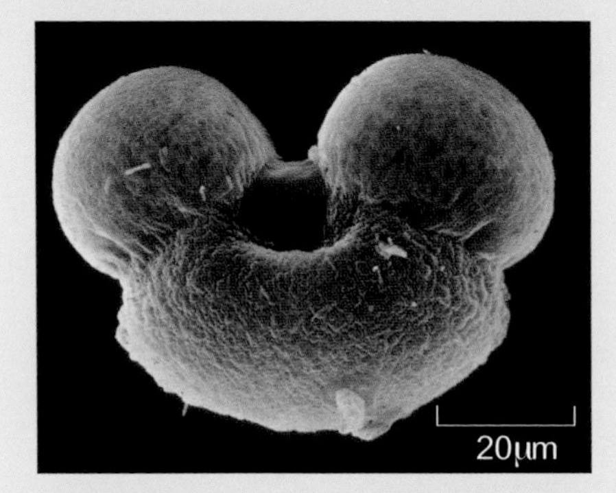


**Figure 5.5**

Pollen from *Pinus* species showing two bulbous air sacs.

*Malva sylvestris*

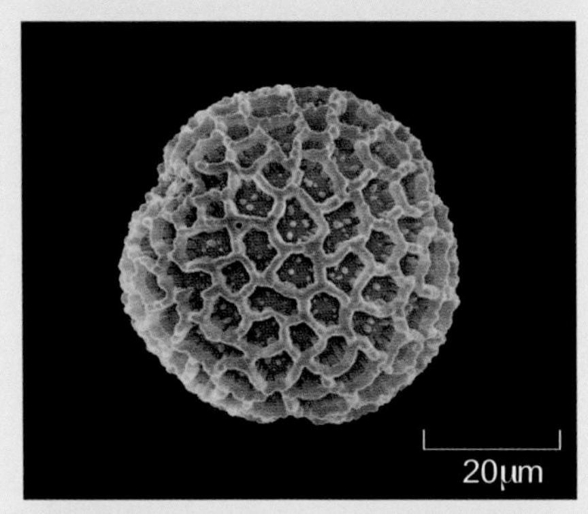


*Delonix leucantha*

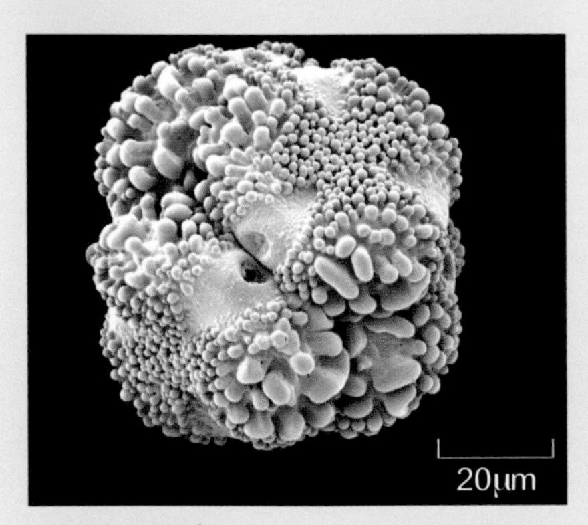


*Dinizia excelsa*

*Gilbertiodendron bilineatum*

**Figure 5.6**

Examples of different pollen structures: common mallow (*Malva sylvestris*) pollen has spines covering its surface, while pollen from the legumes *Delonix leucantha*, *Dinizia excelsa* and *Gilbertiodendron bilineatum* demonstrate reticulate (forming a network), verrucate (lumpy) and striate (stripy) surfaces, respectively.

coarse reticulum. Some have lumps called verrucae, and these can also be variously arranged, and in a variety of sizes, as in the plantains (*Plantago*) and fumitory (*Fumaria*); while others may have coarse stripes as in the edible fruit genus *Prunus*, or very fine ones, as in the maple (*Acer*).

In some families, such as the Rosaceae, pollen structure is remarkably similar for many of the genera. Other families exhibit variations in size, shape, structure and ornamentation: in Acanthaceae, there is so much variation that it is difficult to believe that the pollen grains belong to plants within the same family. Overall, the value of pollen identification to forensic science is variable, depending on the range of pollen sizes and structures exhibited by a particular plant family or genus.

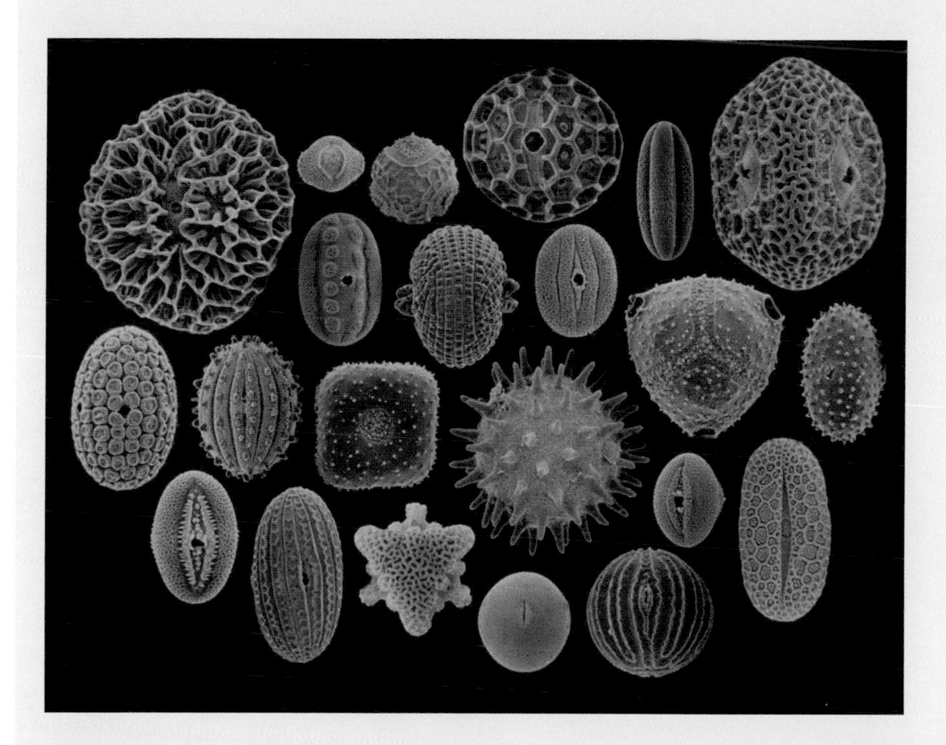

**Figure 5.7**
Pollen grains from various genera of Acanthaceae. (Micrographs and image by Dr Robert Scotland when studying for his PhD at the Natural History Museum, London.)

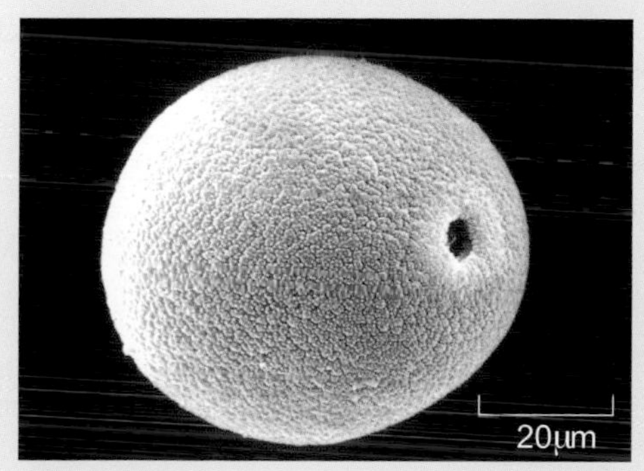


*Poa pratensis*

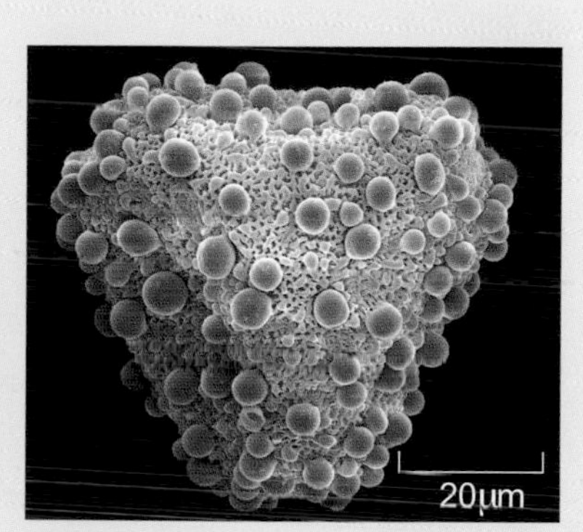


*Eperua*

**Figure 5.8**
The pollen from *Poa pratensis* and other wind-dispersed pollens have a smooth outer coating to aid dispersal and are produced in large quantities by the plants. Pollen that is dispersed by animals, like *Eperua* pollen which is transported by bats, is produced in smaller amounts and can be sculptured to help it adhere to the pollinator.

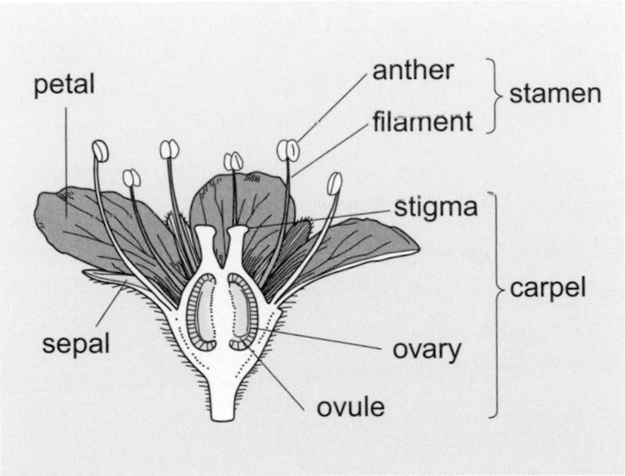

Figure 5.9

The male parts of the flower are the filaments and anthers containing the pollen grains, and the female part is the carpel (stigma, ovary and ovule).

Bryophytes (mosses and liverworts) produce spores that are of limited use as trace evidence due to their relative lack of distinctive features. Exceptions are *Sphagnum* moss and some liverworts. Fern spores are more helpful as many have distinctive scars, sculpturing, or outer coats. Pollen is produced by conifers and flowering plants. It contains the plant's male genetic material and is carried to the female parts of the plant by the wind or by animals such as insects. Pollination results in the fertilisation of an egg contained within the ovule and ovary, which results in the production of a seed.

Spores and pollen grains are microscopic, with most pollen ranging between 20 μm and 70 μm in diameter. However, some large pollen grains, such as those of sunflower (*Helianthus*), can just about be seen by the naked eye. Scanning electron microscopy (SEM) gives beautiful pictures of the outside of pollen grains, but it is of little help in routine forensic palynology. In criminal cases, often as many as 50,000 pollen samples need to be identified and counted during routine analysis using a light microscope.

Wind-pollinated plants produce huge amounts of pollen which can be dispersed considerable distances from the parent plant; where animals are used in distribution, pollination is more targeted. In these plants, the number of pollen grains produced is low and grains rarely find their way into the air because they are carried from flower to flower by the animal (mostly insects). Because of its relative rarity, pollen from insect-pollinated plants can often produce very important and specific trace evidence.

Figure 5.10

The release of pollen from a pine species.

The time of release of spores and pollen depends on the season and on weather conditions. Much pollen, especially that disseminated by wind, floats in the air and then gradually falls as 'pollen rain'. However, the amount of pollen rain that falls is highly variable, both seasonally and from place to place. On rare occasions, the presence of pollen and spores on an object might give information on the time of year of an event.

Biologists interested in allergy studies have created pollen calendars, which record various types of pollen found in the air throughout the year. It is known that in the UK, for example, *Corylus* pollen is typically in the air from December and then tails off towards April, whereas (again in the UK) most pollen from the genus *Pinus* is released in June and tails off towards late summer. However, pollen calendars are very crude and they will vary geographically and from year to year. They have insufficient precision to be used as reliable indicators of time in the forensic context. Furthermore, they do not take into account the fact that some of the pollen and spores that fell to ground in spring will find their way back into the air in windy conditions in autumn. The forensic palynologist needs to have a good grasp of all the factors affecting pollen in the air and pollen rain at any location at any time. Each place (and each crime scene) will have a unique 'palynological signature'.

Pollen and spores can be retrieved from a huge range of objects and materials including, amongst other things, food, clothing, animal coats, footwear, vehicles, rope, soil, dust, carpet, skin, hair, nasal passages and drug resins. Because the pollen rain will be patchy, even within the crime scene, a series of samples must be collected for comparison with items obtained from a suspect. These can then be combined to give a palynological 'picture of place'. The sample used for comparison may consist of soil, water, mud, or vegetation. Indeed, it could be anything likely to have been contacted by the object in question.

The plants represented, and their respective proportions in a sample, make up what is known as the palynological profile of a place or object. These profiles need to be compared one with another. There can never be a perfect match as any item such as a shoe will pick up only a fragment of the palynological profile at a crime scene. Furthermore, the shoe will pick up fragments of other profiles from other places. But, there must be sufficient commonality between the respective profiles for it to be beyond coincidence. Sufficient pollen and spores must be identified and counted for the analyst to be confident of the profile, and this can run to several thousand per sample, but the investigation is always enhanced by the presence of rare components. Ecological rarity is not the same as palynological rarity. Even common plants, such as white clover (*Trifolium repens*) or the harebell (*Campanula rotundifolia*), are palynologically rare by virtue of their low pollen production and insect-pollination.

## Distinctive soil links suspects to murder scene

At the Old Bailey, early in 2009, two defendants were convicted of the shooting of an associate in a quiet, roughly surfaced lane near Romford, Essex, UK. One of the murderers hid behind an oak tree at the edge of the lane and was partially hidden by a cypress hedge. After the shooting, the assailants left in a get-away car, which was later found burned out; they then transferred to another car some distance away. The palynological assemblage found on a balaclava, footwear, and the footwell of the second car, had a high degree of similarity to that of the surface of the lane but not to any of the other places frequented by the suspects.

Furthermore, the profile did not reflect the vegetation of the murder scene. It later transpired that the surface of the road had been made up with soil brought from some other location. This soil was highly distinctive because it must have contained some material from gardens with exotic plants. The same, peculiar assemblage was found in comparison samples, and on the items taken from the suspects. A further link was established in that spores of two rare species of fungus were found on the oak tree trunk, on the hedge foliage, on the balaclava and in footwell of the second car.

## 5.4 Forensic use of plant DNA

DNA analysis is able to identify individual plants or animals with great precision. Some plant DNA databases dealing with within-species variation have been constructed (mostly for economically important plants), but the development of such detailed analyses for most plant species is unlikely to be achieved. The human DNA database in the UK consists of several million individuals, and this covers a single species. There are about 2,000 native plant species growing in Britain alone, so the establishment of comprehensive plant DNA databases is unlikely.

**Figure 5.11**
Palo verde (*Parkinsonia* species) tree (a) and seed pods (b).

a

b

Wherever plant DNA has been used for plant identification, it is mostly at the level of species rather than individual plants. Nevertheless, in 1992, two seed pods of a species of palo verde tree (*Parkinsonia*) were found in a truck seen in the area at the time of the murder of a woman in Arizona, USA. Nine palo verde trees were sampled at the crime scene, as were 19 from other places, but the DNA profile of the pods matched only that of a single tree growing at the crime scene, a match that had just a one in 136,000 chance of occurring at random.

Unlike humans, plant species populations can consist entirely of clones and these clones can cover huge areas. Plant clones arise from plants that have been vegetatively propagated from one another and so their DNA profile will be identical (you will learn more about this in Chapter 9). Furthermore, commercial cloning through, for example, cuttings and grafts, has resulted in individuals of a single clone being distributed to many places throughout the country. Thus, at a crime scene, it might be difficult to establish that a particular plant was definitely the source of specific trace evidence. For most plant species, many individual plants would be present at the scene and it would be impractical to test enough plants to be sure that the source of the matching evidence was included in the testing.

Sophisticated analytical techniques, including DNA analysis, have dominated forensic science in recent years, but there are significant numbers of cases in which these techniques fail to provide adequate evidence for the prosecution. In a number of criminal investigations since around 1995, plants have provided the *only* forensic evidence and this has been accepted by the courts to aid the prosecution's case.

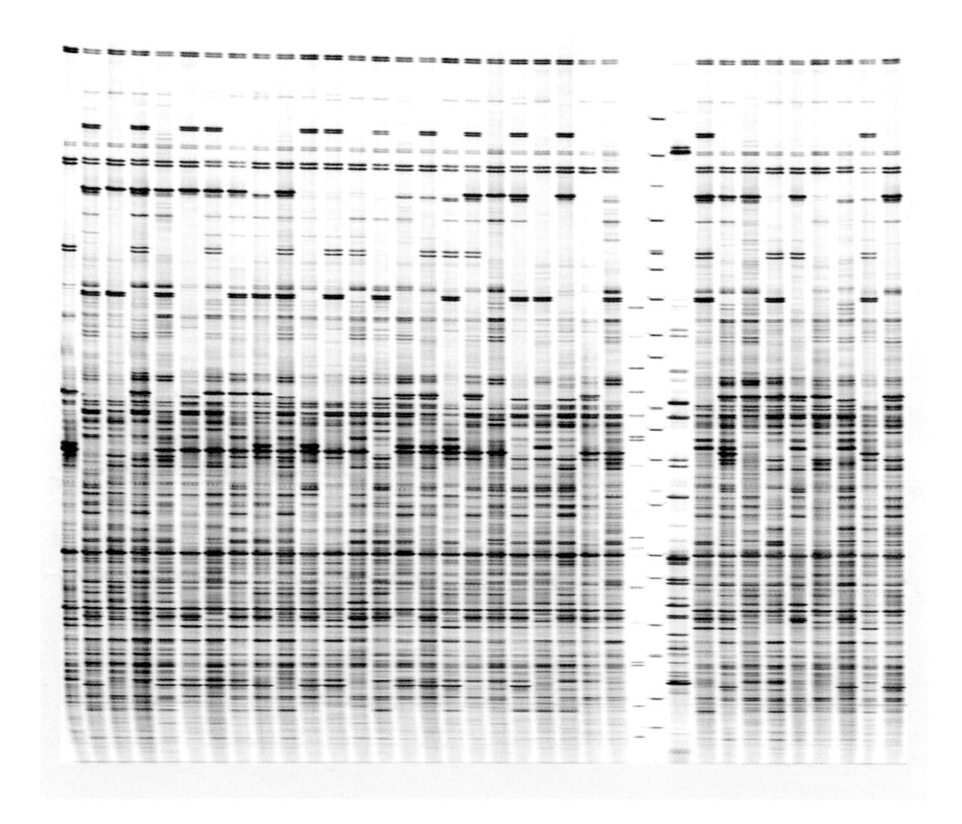

**Figure 5.12**
A DNA profile from maize (*Zea mays*). Each column represents the DNA from an individual plant and the differences between individual plants can be seen at the points where bands are present in one column but missing in others.

# 6 Plants for nutrition and well-being

Plants are autotrophic (self-feeding) organisms; they capture atmospheric carbon dioxide and convert it into carbon compounds, which they use to create the range of substances needed for growth and development.

Animals, including humans, are heterotrophic and cannot make all the essential chemicals that are necessary for their survival. Heterotrophic organisms, therefore, have to obtain their nutrients either directly from plants (herbivores) or from other animals lower down the food chain than themselves (carnivores) or from a mixture of both (omnivores). This chapter introduces the nutritional importance of plants to humans, who are largely omnivores, as well as describing the role of chemicals in the creation of aroma, flavour and colour in plants, all of which contribute to our general well-being. The difficulty in writing one chapter covering these areas is that much must be left out, so read the following as an 'appetiser'!

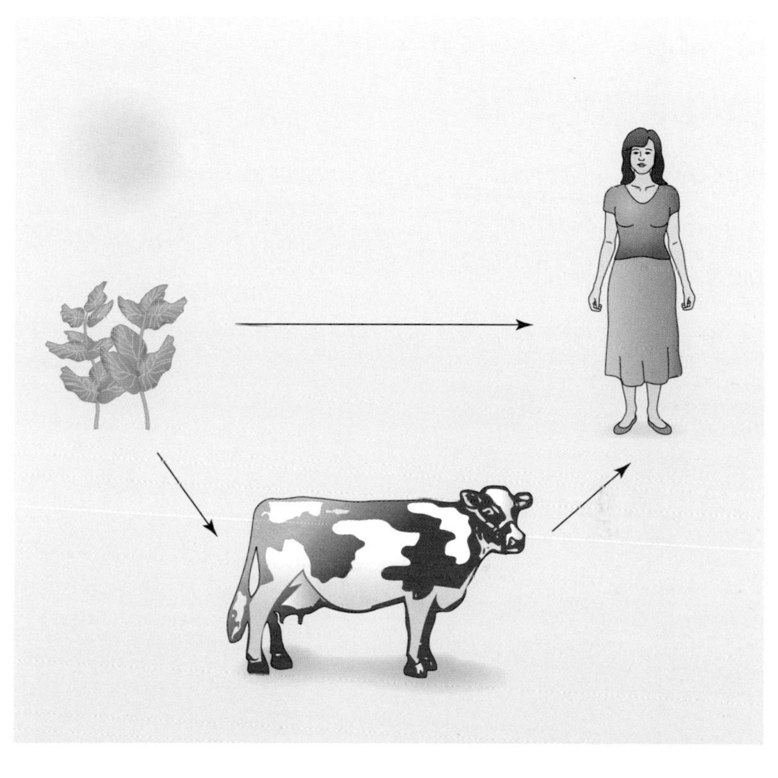

**Figure 6.1**
A simple food chain showing how heterotrophic organisms, such as humans and cows, depend on autotrophic organisms like plants. Note the arrow symbol means 'Is eaten by'.

## 6.1 Components of the human diet

A healthy human diet consists of water and five main types of essential nutrients: those needed in large quantities, the so-called macronutrients (carbohydrates, fats and proteins) and those needed in small amounts, known as micronutrients (vitamins and minerals). A range of different food groups can provide these nutrients.

Human diets that are rich in plants and plant products, such as fruit and vegetables, are considered to be the healthiest, and the recommendation is to eat at least five portions of fruit and vegetables every day. Fruit and vegetables are a valuable source of vitamins and minerals. Plants also provide us with starch (a valuable energy source) and other carbohydrates, found mainly in seeds, grains and tubers. Pulses provide us with proteins, and the rich variety of plant oils contribute to the fats needed in a healthy diet.

**Figure 6.2**

In recent years, governments have been providing people with dietary advice. The UK's Food Standards Agency has produced the eatwell plate to show the relative proportions of the different food groups that should be consumed in order to have a healthy diet. Around two-thirds of human daily intake should be from plant sources: one-third from fruit and vegetables and one-third from bread, cereals and potatoes. The foods containing a high proportion of fat and sugar, which are part of the remaining one-third, should only be consumed in small amounts. These can also originate from plants; even the proteins and dairy products from animals would not exist without plants.

## Micronutrients

Vitamins and minerals are needed for a range of bodily functions, including for metabolic processes and for communication between cells. Plants are a good source of vitamins and minerals which cannot be synthesised within the human body and so must be obtained from our diet.

| plant source | | | | |
|---|---|---|---|---|
| fat-soluble vitamins | | E | A, E | E |
| water-soluble vitamins | $B_1$, C, folate | $B_1$, C, folate | | |

**Figure 6.3**

The main sources of fat-soluble and water-soluble vitamins derived from plants. All vitamins, except vitamin D and $B_{12}$, can be obtained directly from plant sources. Vitamin A can be synthesised in the body from β-carotene, which is found in dark green vegetables and carrots. Vitamin D can also be synthesised by the body but sunlight is required for this process. Vegetarians and vegans need to take care to ensure that they eat fortified cereals to get an adequate intake of vitamin $B_{12}$. Water-soluble vitamins, such as the vitamin B group and vitamin C, cannot be stored in the body as they are excreted in the urine and need to be replenished frequently. By contrast, fat-soluble vitamins, such as E, D and K, can be stored within fatty tissue. Vitamins, particularly A, C and E, have also been shown to play an important role by acting as antioxidants.

## What are antioxidants and why are they important?

During chemical reactions within the body, harmful chemicals are produced which are highly reactive and unstable and can cause damage to cells, leading in extreme cases to cancer. These damaging chemicals are called free radicals. Antioxidants are chemicals that have the ability to 'mop-up' and make these free radicals less reactive, hence reducing cellular damage. Free radicals are often implicated in the ageing process and may be involved in diseases such as cancer, heart disease, rheumatoid arthritis and Alzheimer's disease.

Several vitamins, notably vitamin C, act as antioxidants, as do many of the compounds that give plants their colours. Many of these compounds are part of the flavonoid group (see p. 72) and their presence in berries such as cranberries and blueberries has led to these foods being classed as 'superfoods'.

Some diseases are caused by deficiencies in vitamins and minerals. Such deficiencies are now rare in the developed world but when they do occur are most commonly seen in the water-soluble vitamin group. The disease beriberi (see p. 20) is caused by a diet deficient in thiamine (vitamin $B_1$), which is found in the husk of rice. Folate or folic acid is essential for the synthesis of DNA, and folate deficiency can lead to anaemia (lack of haemoglobin in red blood cells) because the cells cannot divide and multiply at a high enough rate. Pregnant women are advised to take folate supplements to reduce the chance of the fetus developing spina bifida ('split spine').

## Vitamin C and scurvy

One of the best understood deficiencies is that of vitamin C, which leads to the disease scurvy. In the sixteenth and seventeenth centuries many sailors aboard naval vessels that were at sea for prolonged periods suffered from scurvy. The disease has a wide range of symptoms including gum disease, anaemia and muscle pains.

Eventually death could result: the explorer Vasco de Gama lost more than half his crew from the disease on his trip around the Cape of Good Hope in 1497. The British navy, in particular, were keen to understand how to treat scurvy and in 1747 a Scottish Naval Surgeon, James Lind, successfully treated sailors suffering from the disease by giving them lemons and oranges. It is now known that all citrus fruit contain vitamin C, which is required in the body for making collagen, a connective tissue, as well as having a role in wound healing and in allowing the body to absorb iron. The subsequent Royal Navy practice of giving citrus fruit, including limes, to sailors resulted in British sailors being given the nickname 'limeys' by their American counterparts, a term that has since spread to now encompass all British people.

**Figure 6.4**

A painting of James Lind treating sailors with scurvy on board HMS *Salisbury* in 1747.

## Carbohydrates

Carbohydrates are compounds made from carbon, hydrogen and oxygen, such as sugars and starches. They provide us with the energy needed for all the biochemical processes and physical activities necessary during every moment of our lives. This energy is released from sugar by the process of cellular respiration in all living cells.

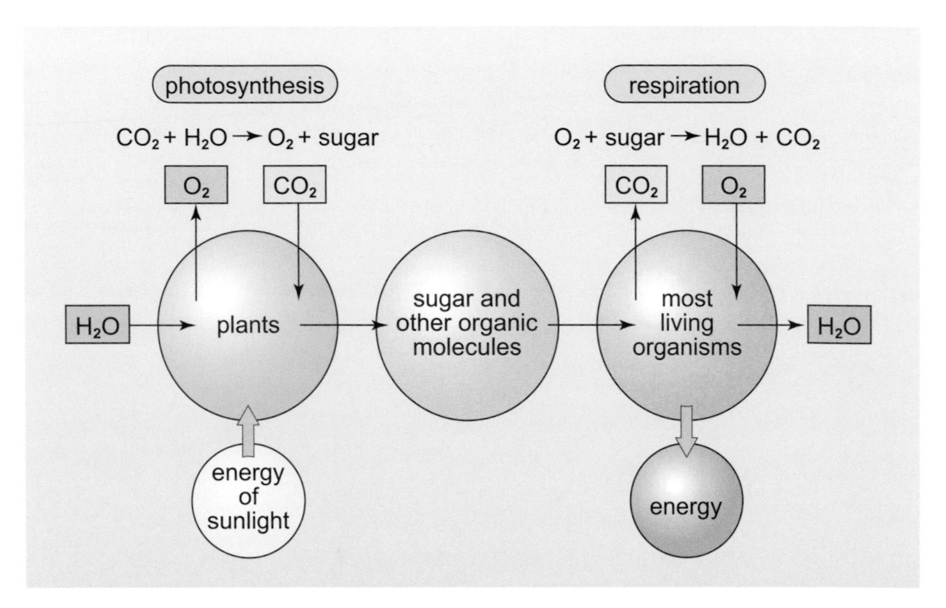


**Figure 6.5**
The process of cellular respiration is essentially the reverse of photosynthesis but note that plants both respire and photosynthesise.

Carbohydrates are a complex group of polymers built from sugar monomers such as glucose. The way the sugar molecules are linked together determines the type of polymer that is formed. As Chapter 3 (p. 37) has shown, starch is a polymer of glucose, and it can be broken down during digestion to release individual glucose molecules that can be used in cellular respiration. Cellulose, however, is a glucose polymer with a structure that cannot be digested by humans. Cellulose and other similar indigestible carbohydrates form a crucial part of our diet in the form of fibre. Dietary fibre is essential for the efficient and healthy passage of food through the alimentary canal and for effective bowel movement. This is because the fibre results in bulkier gut contents and this, in turn, stimulates gut muscle activity causing a more rapid transit of fibre-rich food. There is some evidence that a diet rich in whole grains, fruit and vegetables, and thus rich in fibre, helps prevent bowel cancer as toxic waste food products remain in the bowel for shorter periods. Additional evidence suggests that fibre may actually bind to the potentially toxic substances and therefore aid their elimination from the gut.

## Fats and lipids

The term lipids refers to the group of compounds containing fats, oils, waxes, cutin (found on leaves) and tuberin (found on cork). They are used as energy storage molecules, to form cell membranes and may have structural roles. Lipids are made from glycerol and fatty acids. Plant lipids commonly contain $\alpha$-linolenic acid (an omega-3 fatty acid) and linoleic acid (an omega-6 fatty acid), both essential in the human diet as our bodies cannot make them. Seeds and nuts have a high concentration of both saturated fatty acids, such as palmitic acid, and unsaturated fatty acids, such as linoleic acid. A small amount of oils (or fats) is required in the human diet, although it is generally considered healthier to have a higher proportion of unsaturated fats rather than saturated fats. Lipids play important roles in the human body: they are crucial in forming cell membranes; they act as energy storage molecules; they provide insulation for internal organs and they enable fat-soluble vitamins to be absorbed and stored.

## Proteins

Proteins are essential for growth, maintenance and repair in both animal and plant cells. They contain nitrogen, as well as carbon, hydrogen and oxygen. Plants therefore require a source of nitrogen in far greater concentrations than any other inorganic nutrient. They obtain nitrogen from the soil naturally via the nitrogen cycle.

### The nitrogen cycle

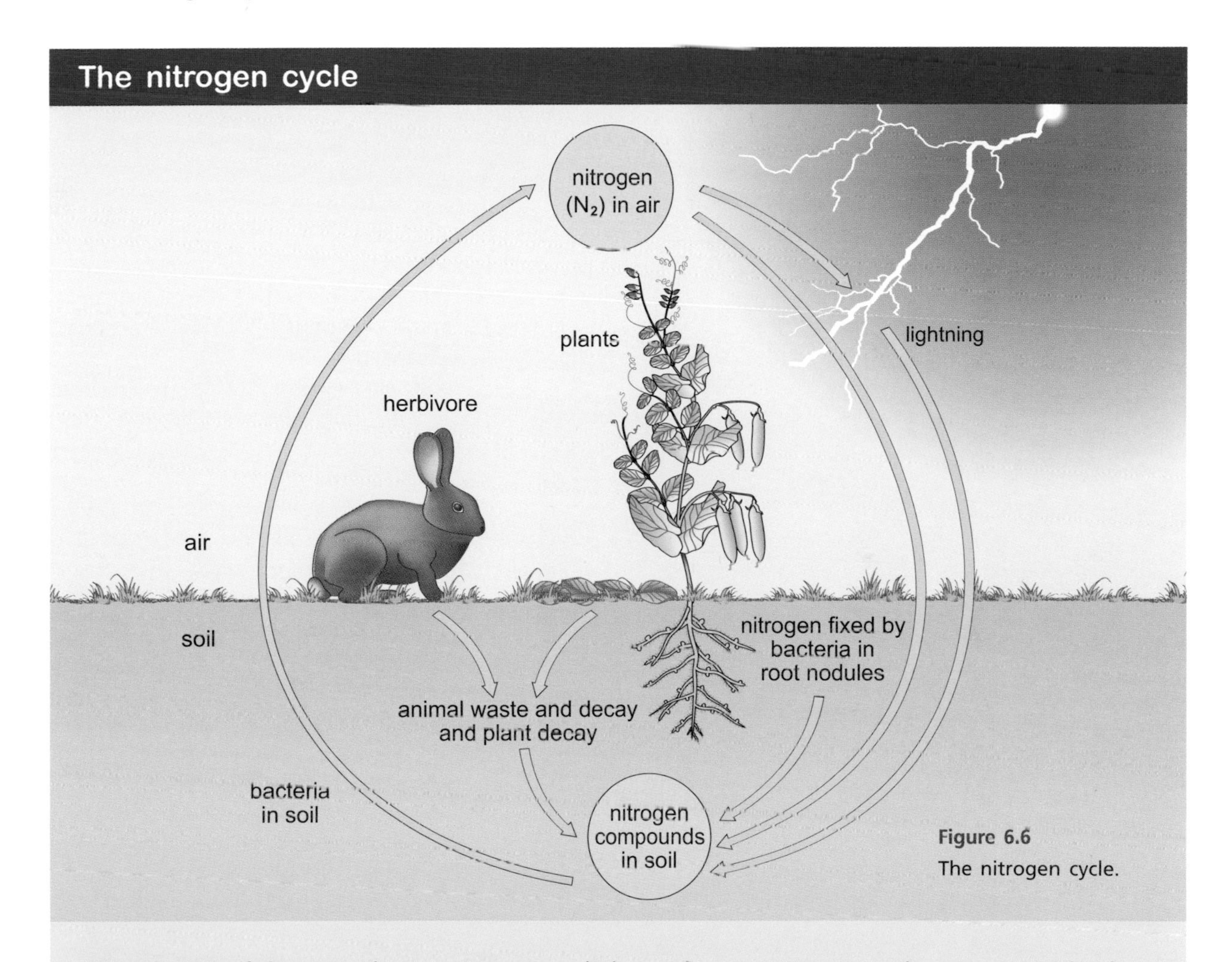


**Figure 6.6**
The nitrogen cycle.

Although 78% of the atmosphere is nitrogen gas ($N_2$), very few organisms can utilise nitrogen in this form. However, there are some soil bacteria which can convert nitrogen into nitrates, and these nitrates can be absorbed by plant roots and used to form plant proteins. When plants are eaten, the plant proteins get passed on to animals. The animals break them down and then re-assemble the components into nitrogen-containing animal proteins. Waste nitrogen can be excreted from animals in the form of urea and uric acid in the urine, and the nitrogen is once again available for plants to take up.

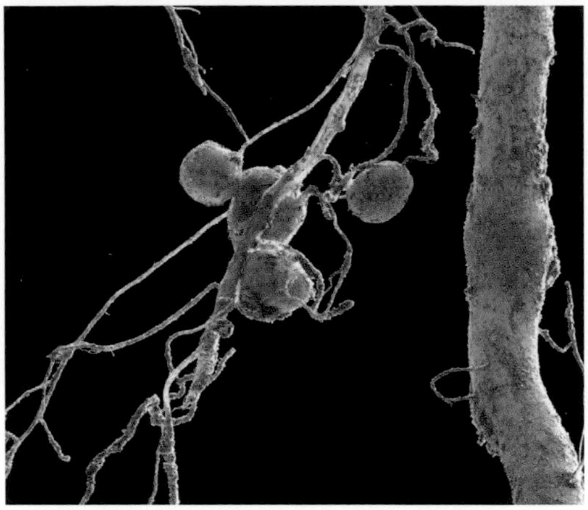

Some plants, such as the Leguminosae (also known as Fabaceae) family (peas, beans and clover) and alder (*Alnus glutinosa*), can actually 'fix' their own nitrogen by forming an association with bacteria present in nodules on their roots. These bacteria can capture atmospheric nitrogen and convert it into forms such as nitrate, which can subsequently be used by plants.

**Figure 6.7**
Nitrogen-fixing root nodules of the common bean (*Phaseolus vulgaris*).

## 6.2 Plant flavours and aromas

The flavours and aromas of foods are due to the volatile nature of the chemicals within them, which are released from the food in gaseous form. Generally speaking, the flavours we experience are due to us smelling these chemicals rather than through tasting the food. Volatile chemicals can be released from plants in a variety of ways: as fruit ripens, when leaves of herbs are bruised or chopped, when food is cooked, or as the food is chewed with saliva and mechanically and enzymatically digested in the mouth. Not only do these flavoursome chemicals and compounds provide enjoyment when food is eaten, many also have health benefits.

**Figure 6.8**
The pleasure of food is as much due to the aroma as to the taste.

A major group of chemicals that provide flavour are the terpenoids, also called isoprenoids, which are a wide and diverse group of nearly 23,000 identified compounds. Terpenoids are responsible for the aroma of ginger, cinnamon and cloves. There are six main subdivisions of the terpenoids, ranging from the chemically simple monoterpenoids and sesquiterpenoids, which provide scent and flavour, to the extremely complex carotenoids, which provide colour. Terpenoids are polymers made up of chains of a monomer known as the 'isoprene' unit.

Within the plant, terpenoids play very diverse and contradictory roles, both in defence of the plant and as attractants, encouraging animals to eat the fruit so that the seed will then be dispersed via excretion. Monoterpenoids are found extensively in the leaves of many aromatic plants and are associated with aromatic herbs such as mint, and strong flavours such as lemon and orange.

The plant defence properties of terpenoids have been commercially exploited for pest control. For example, citronella candles can be burnt to ward off unwanted flying and biting insects. Citronella is extracted from the

**Figure 6.9**
*Mentha spicata* (spearmint).

leaves and stems of various members of the lemongrass (*Cymbopogon*) genus and is used extensively in the perfume industry. Lemongrass is also used to make tea, soups and curries and imparts a distinctive citrus flavour. Trials have shown that lemongrass oil can ward off mosquitoes, but more research is needed to extend the period of time for which it is effective.

**Figure 6.10**
Citronella candles are burnt to release terpenoid-based vapours, which repel many flying insects.

a

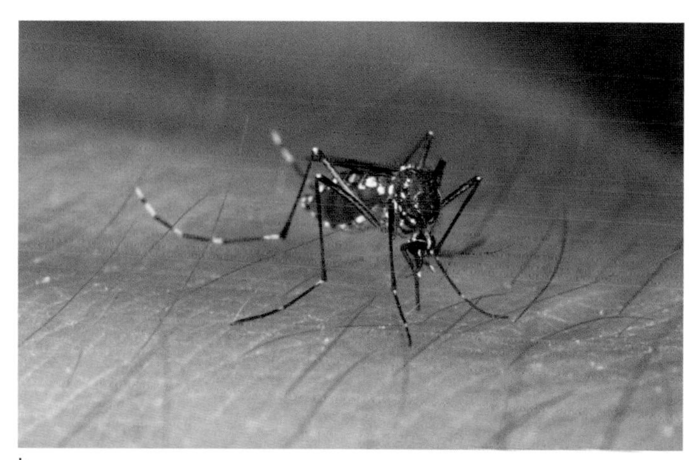

b

**Figure 6.11**
Oil extracted from the lemongrass plant (a) is a promising avenue of research in the search for effective natural repellents of insects such as mosquitoes (b).

## Sage

Both common sage (*Salvia officinalis*) and Spanish sage (*Salvia lavandulifolia*) are popular culinary herbs and for centuries have been used for their medicinal properties. Indeed, the name *Salvia* comes from the Latin word meaning 'to heal'. Sages have been used traditionally to treat a whole array of ailments, including muscle spasms and tension, as well as for their antibiotic and antifungal properties. Recent research, however, has supported the belief that sage can help manage mild and moderate cases of Alzheimer's disease. Although sage contains a cocktail of different chemicals, it is believed that monoterpenoids play an essential role in helping relieve the symptoms of this debilitating disease. Other culinary herbs such as rosemary (*Rosmarinus officinalis*) and lemon balm (*Melissa officinalis*) are being investigated for similar properties.

**Figure 6.12**
Common sage (*Salvia officinalis*).

Not all flavours derive from monoterpenoids or sesquiterpenoids. Other more pungent flavours are the result of chemicals containing sulphur. Such chemicals are found in strongly tasting vegetables like the alliums (Amaryllidaceae), which include onions, chives and garlic, and from leafy vegetables like the brassicas (Cruciferae/Brassicaceae), which include brussel sprouts, broccoli, cabbage and mustard. Consumption of alliums has been reported to have beneficial effects in preventing oesophageal, stomach and colorectal cancer.

There are over 120 known types of glucosinolate, organic compounds that contain both sulphur and nitrogen, and their role in plants is primarily a defence against herbivores. Glucosinolates are one half of a two-component system that is activated by tissue breakdown during mechanical damage (for example, by chewing) and by chemical breakdown within the gut. The second component is an enzyme known as myrosinase, which comes into contact with the glucosinolate during tissue breakdown. As a result an unstable compound is formed resulting in the pungent tasting chemicals that are commonly experienced when eating 'greens', sometimes known as the 'mustard bomb'. The benefit of glucosinolates and their derivatives in the human diet is their association with the prevention of cancer, especially pancreatic cancer, and heart disease.

**Figure 6.13**
Vegetables known to contain high levels of sulphur-containing glucosinolates.

## 6.3 Plant colours

Besides the chemicals that provide flavour and scent, many plant tissues including leaves, flowers and fruits harbour a whole array of chemicals that are beneficial to human health and nutrition. One such group of chemicals gives rise to colour, which is important for making food look attractive and appetising and also for attracting pollinating insects to flowers.

### Carotenoids

Figure 6.14 - left
Many fruits and vegetables are extremely colourful: this is entirely due to the many pigments they contain.

Figure 6.15 - right
Carotenoids give some flowers and fruit their bright colours that attract pollinating insects and birds, which eat the fruit and spread their seeds.

Carotenoids are a major group of pigments that give yellow, orange and red colours to a wide range of leaves, flowers, fruits and roots. Just as chlorophyll absorbs most of the colours in white light except green, which is reflected, the carotenoids absorb the blue end of the visible light spectrum and reflect the yellow to red light.

Carotenoids are divided into two sub-groups, those containing oxygen (xanthophylls) and those that are oxygen-free (carotenes). They have a variety of important functions, including playing a crucial role in photosynthesis and mopping up free radicals in plants; in this latter function carotenoids play, for plants, a similar role to that of vitamins in the human diet. The carotenoids are also crucial as pigments, providing bright flowers that attract pollinators and colourful fruits that are devoured by animals, which then spread the seed.

About 50 of the 600 known carotenoids are normally eaten in the human diet and many are thought to provide health benefits, particularly in preventing certain diseases and cancers. The role of carotenoids as antioxidants has been studied in great detail, and the beneficial roles of the pigments β-carotene (orange), lycopene (red), lutein (bright yellow) and zeaxanthin (yellow) are well documented.

Tomatoes contain more lycopene than any other fruit or vegetable. However, processed tomato products, such as paste and sauces, are considered a better source of dietary lycopene because the pigment can be absorbed more efficiently into the digestive system from these products than from fresh tomatoes. Lycopene has been shown to play an important role in the prevention of prostate cancer.

Lutein and zeaxanthin have been shown to have a protective role against eye diseases as these pigments absorb and dissipate the potentially damaging blue and ultraviolet light. Spinach is a particularly rich source of lutein, as are other dark green leafy vegetables and many yellow flowers, including marigolds.

Carrots contain high levels of β-carotene. This orange pigment is of particular interest because it can be converted to provitamin A, which is required in the human diet to prevent vitamin A deficiency. People who lack β-carotene in their diets and suffer from vitamin A deficiency are at risk from blindness and an impaired immune system.

## Flavonoids

Flavonoids are a second major group of plant chemicals and include pigments as well as structural compounds like lignin. Some flavonoids give rise to plant defence chemicals such as tannins. The production and concentration of flavonoids is very dependent on the environment, health and growth stage of the plant. Many fruits and vegetables contain high levels and mixtures of flavonoids, but the chemicals can also be induced in plants by stress and as the result of wounding and attack.

Along with carotenoids, three classes of flavonoids – anthocyanins, flavones and flavanols – form the basis of many of the bright colours commonly found in fruits, flowers, stems and roots. Reds, purples and blue colours are attributed to the anthocyanins.

**Figure 6.16**

The three most common anthocyanin pigments, pelargonidin, cyanidin and delphinidin, were originally extracted from flowers of pelargonium, blue cornflower and delphinium, respectively.

Pelargoniums contain pelargonidin

Blue cornflowers contain cyanidin

Delphiniums contain delphinidin

**Figure 6.17 - below**

The bright red autumn leaf colours, particularly associated with *Acer* species, are due to anthocyanin accumulating on sunny, cool autumn days as the green chlorophyll pigments are withdrawn from the leaves. Exactly why anthocyanins accumulate in leaves in some species is still the subject of scientific research.

Figure 6.18
Anthocyanins are commonly found in red and purple foods, such as red cabbage, blueberries, blackberries and blackcurrants.

Figure 6.19
The flavones and flavanols are responsible for the yellow, cream and white pigments associated with pale flowers and fruit, from left to right: snowdrops, white iris and white currants.

Tannins provide a defence mechanism within the plant by being unpleasant to consume due to their dry, astringent properties. The term was first introduced in 1321 to describe the chemicals extracted from plants that were used to tan leather. Tannins also give rise to the brown colours associated with autumn leaves and the dark colour of black tea.

Like carotenoids, many flavonoids are antioxidants. Fruit and vegetables contain a whole array of different flavonoids, not just those associated with colour. The anthocyanins have been of great interest to the media, leading to the term 'superberries' being applied to blueberries, blackcurrants and black grapes. The media have also made much of the health benefits gained from drinking red wine. Studies have shown that the darker, heavier red wines, such as Bordeaux and Shiraz, contain higher levels of beneficial flavonoids; therefore there could well be benefit from drinking a small glass of red wine a day!

栏
长年为您
产中药材

# 7 Medicinal plants

Around 50,000 plant species, that is about 1 in 6 of all known species, have recorded medicinal uses. Plants continue to be the basis of important new pharmaceutical drugs, are widely used for self-medication in western medical systems, and are the main form of medicine in the developing world. The World Heath Organisation estimates that about 80% of the world's population still rely on plants for their primary source of medicines.

## 7.1 History of medicinal plants

The early history of herbal medicines is obscure. It can be hard to tell if plant remains found in archaeological sites were used for medicinal or nutritional purposes. In ancient texts, plant names (and the nature of the illnesses treated) are hard to translate. For example, it has proved possible to identify only a quarter of the 160 plant products mentioned in ancient Egyptian medical texts.

The best evidence for the great antiquity of herbal medicine comes from the study of animal self-medication. A study of monkeys in eastern Brazil found that they consumed an unusually high proportion of tannin-rich plants that are effective against intestinal parasites. The monkeys are virtually parasite-free. Great apes have been seen to collect over 30 plant species with medicinal properties, and similar behaviour has been noted in many other animals, including elephants, rhinos, wild boar, bears and birds. It is likely that such a widespread practice among the primates would have formed part of the behaviour of the first humans. Observation of animal self-medication may also have guided the first experiments in the use of plants by humans.

The earliest records of plant medicine are on the clay tablets of the Sumerians (3000–2000 BC) who lived in Mesopotamia. The best-studied are the medical papyri of ancient Egypt, including the Edwin Smith papyrus and the Ebers papyrus, dating to 1500 BC and possibly deriving from the tomb of a doctor.

**Figure 7.1**
Part of the Ebers papyrus, named after the German Egyptologist who purchased the scroll. The information on the scroll includes a potential remedy for asthma which was 'a mixture of herbs heated on a brick so that the sufferer could inhale their fumes'.

The Ayurvedic medical system of India probably reached a recognisable form about 2,450 years ago, and is documented in Sanskrit manuscripts dating from AD 400 onwards. The development of Ayurvedic medicines was influenced greatly by the exchange of plants through Arab traders who, for example, are thought to have introduced the use of opium into India for the treatment of conditions such as dysentery.

The earliest medical manuscripts of China date back to 2,000 years ago. Many of these are referred to in Li Shizhen's *Materia medica*, a sixteenth-century encyclopaedia that consisted of 52 volumes, covering 1,892 medicinal substances of which 60% were plants. Traditional Chinese medicine (TCM) was formalised in its current form under Chairman Mao Zedong in the 1950s, but incorporates many elements from the preceding 2,000 years of practice.

## 7.2 Herbal treatments in medicine

**Figure 7.2**
Lungwort (*Pulmonaria officinalis*) was used to treat respiratory diseases. The plant contains tannins and other compounds that have an expectorant, soothing effect.

The way a society uses herbal medicines relates to its wider beliefs about medicine. The Hippocratic system, developed in ancient Greece by 400 BC, remained the basis of most European medicine until the eighteenth century. Illness was seen as the result of an imbalance in the body of the four humours (fluids), which were bile, phlegm, blood and black bile. Plants were believed to have the property of being cold, moist, hot or dry, and could thus correct the imbalance. The shape of a plant was also thought to indicate its medicinal properties. For example, plants with lung-shaped leaves could be used to treat respiratory problems.

In western medicine, herbal medicines are usually made from purified plant extracts, often containing one or a few compounds. By contrast, in traditional medicine treatments, extracts or tinctures are usually made from an entire plant part so will contain many compounds. Mixtures of different plants are also often used; it is argued that the resulting mixture of plant compounds leads to safer medicines, with some compounds acting as buffers to any that cause damage.

In many societies there is a distinction between formal, written systems of herbal medicine, and folk medicine passed on within families by word-of-mouth, often in rural areas. Folk medicines are harder to study, as written evidence relating to them is scarce. It is interesting to note that the formal system of TCM uses about 500–600 plant species, whereas folk medicine in China uses about 10,000 species.

In early nineteenth-century Europe, chemists developed techniques for the analysis and isolation of plant compounds. In 1805 morphine, a very effective pain reliever or analgesic, was isolated from the opium poppy, and in 1820, quinine was isolated from the bark of the fever bark tree (*Cinchona calisaya*) as a treatment for the symptoms of malaria.

a

b

Figure 7.3

(a) The fever bark tree (*Cinchona calisaya*) and (b) its bark, which was used to treat symptoms of malaria.

a

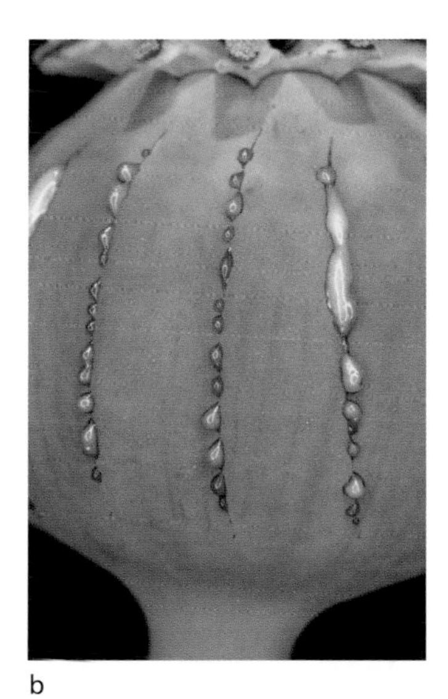

b

Figure 7.4

(a) The opium poppy flower (*Papaver somniferum*) and its seed capsule, from which morphine can be extracted.

(b) To extract the morphine, immature green opium seed capsules are cut so the milky sap within is released. The sap is collected and then dried.

## Plant alkaloids

Morphine is an example of an alkaloid, a diverse group of chemicals found in plants. Humans have used plant-derived alkaloids: as drugs, in the case of the anti-malarial compound quinine; as stimulants such as nicotine, one of the active ingredients in tobacco (*Nicotiana tabacum*); and as poisons in the examples of scopolamine and hyoscyamine, which come from the deadly nightshade (*Atropa belladonna*). Deadly nightshade also contains another alkaloid, atropine, which is toxic when ingested but can be used as a drug in eye surgery as it causes the pupils of the eye to dilate.

In 1820, plant-based drugs accounted for 70% of the contents of the *American Pharmacopoeia* (the standard official authoritative text for pharmaceutical substances and medicinal products in America); by 1960 the proportion had fallen to 5%. In the intervening 140 years, developments in organic chemistry led to the synthesis of many drugs. An excellent example of a synthetic chemical being produced to mimic the effect of a naturally occurring, plant-derived chemical is the drug aspirin.

# 7.3 The story of aspirin

## The discovery of aspirin

It is tempting to speculate that, in the search for relief from pain or fever, our ancestors tried out the plants that grew around them. There are hints that the link between willow and the relief of pain was established many thousands of years ago. The earliest detailed record alluding to the use of willow (*Salix* species) is in an ancient Egyptian papyrus written in 1534 BC that suggests treatments for various illnesses. It recommends willow be taken by mouth as a tonic and also as an external salve for ear infections and sore muscles. The willow was to be used with other plant products, such as figs, beer and dates. Hippocrates, who is still regarded as the 'father of medicine', listed willow extract for relieving labour pains in about 400 BC.

In 1758, an English clergyman, Edward Stone, nibbled the bark on a willow twig and was reminded, because of the bitter taste, of the fever bark tree (*Cinchona calisaya*), which was already used to treat the symptoms of malaria. He intuitively guessed that the two trees might contain a similar compound, so he collected willow twigs and got the local baker to dry them. He then ground them into a powder with which he conducted a series of clinical trials on the poor and needy of Chipping Norton, and on members of his family and servants. He varied the doses and kept records of the results and noted that it was only effective for certain fevers. After five years of trials, he wrote up his findings for the Royal Society in London and his paper was published in their journal *Philosophical Transactions* in 1763.

## Analysis and isolation of the active ingredient

In 1828, Johann Buchner from Munich extracted from willow bark small quantities of bitter-tasting yellow crystals, which he called salicin (from the Latin name *Salix*). A French chemist called Henri Leroux improved the extraction technique, but an Italian named Raffaele Piria extracted an even more potent substance which he called salicylic acid. Treatment of patients with pure salicylic acid was started but a side effect of stomach ache was noticed, especially if the drug was given in large quantities. Eventually a substance, acetylsalicylic acid was produced that was effective for the treatment of pain but caused fewer side effects. By the end of the nineteenth century, large-scale production of pure aspirin (the trade name given to the product by the Bayer chemical company) was underway.

**Figure 7.5**
Willow trees (*Salix* species) have bark that contains the active ingredient in aspirin.

## The effects of aspirin

There have been many claims and counterclaims for the effect of aspirin in the body. There is general agreement that aspirin acts as an:

- analgesic, reducing pain
- antipyretic, reducing fever
- anti-clotting agent in the blood
- anti-inflammatory.

Until the advent of two other pain-relieving drugs, paracetamol in the 1950s and ibuprofen in the 1960s, sales of aspirin were very high: in fact they contributed a third of the total turnover of the Bayer chemical company. Between the 1960s and the late 1990s sales of aspirin declined, but since then sales have increased again due to its prescription as a preventative medicine for patients with coronary heart disease who are at high risk of a heart attack. Aspirin is still heavily used throughout the world and it is estimated that 120 billion aspirin tablets, equating to 40,000 metric tonnes, are consumed each year.

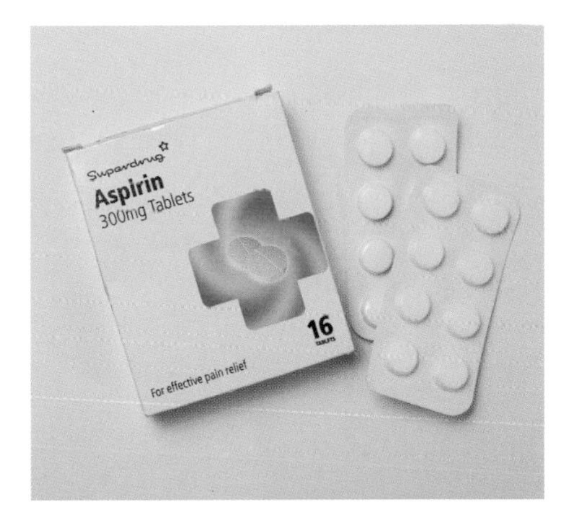


**Figure 7.6**

Aspirin tablets are one of the most widely used drugs in the world.

**Figure 7.7**

Although, overall, fewer drugs are now based on natural products, in some areas of medicine the use of plant-derived drugs is high. For example, over 60% of the modern drugs being used to treat cancer are based on natural compounds. Late twentieth-century successes include vinblastine and vincristine, derived from the Madagascar periwinkle *Catharanthus roseus*, for treating childhood leukaemia (a form of cancer).

## 7.4 The story of taxol

The discovery, characterisation and subsequent chemical synthesis of taxol is a classic example of how modern drugs based on naturally occurring plant-derived compounds can be developed and found to have a highly effective and novel mode of action.

### The discovery of taxol

The story of taxol, also known as paclitaxel, began in 1960 when the National Cancer Institute (NCI) in the USA organised botanists from the Department of Agriculture to collect and screen plants as possible sources of anti-cancer drugs.

In 1962, in the third year of collecting, Arthur Barclay and his students were in Washington State on the west coast of the USA. They had collected specimens from various herbaceous and shrubby plants when they encountered the Pacific yew tree (*Taxus brevifolia*). This tree grew among the magnificent specimens of redwoods in the dense forest of the Pacific Northwest. In the shade of the canopy trees, Pacific yew did not grow very tall and was not regarded by the local woodsmen as being of any value to the timber trade. Two collections were made and sent to several laboratories for screening: one of twigs, leaves and fruit and the other of bark. Extract from the bark showed the ability to be toxic against animal cells, so-called cytotoxic activity, and further samples of bark were collected and investigated to isolate the active toxic compound.

**Figure 7.8**
The Pacific yew tree (*Taxus brevifolia*).

Analysing and separating the many compounds contained in plant cells is a very complicated process and Dr Monroe Wall, a skilled scientist, led the research team. Their research revealed that the yew bark extract was effective against mouse leukaemia and the task of purifying the active ingredient gathered momentum. By repeatedly separating the components of the extract and testing each one for activity against cancer cells, they finally identified the single substance responsible for the anti-cancer activity in 1966. This substance was named taxol, linking it with the yew's name *Taxus*. The chemical structure of taxol was finally elucidated in 1971. Research on the effectiveness of taxol against cancerous tumours and leukaemia has continued, and it is now an effective anti-cancer drug, used to shrink skin cancers, as well as ovarian and breast cancers.

### How taxol works in animal cells

The nuclei of multicellular organisms, such as plants and animals, divide during the natural cellular reproduction process to allow for growth or to produce replacement tissue. If this nuclear division is not controlled, a mass of cells is produced which is referred to as a tumour. In 1977, Dr Susan Horwitz, from the Albert Einstein Medical School in New York, was asked to investigate taxol to find out how it stopped tumours growing. She had been invited to do this work because of her interest in small molecules that are biologically active.

In 1979, Horwitz and her team announced they had found that the taxol molecules prevented cells dividing and actually led to cell death, which in turn led to the shrinking of tumours. This announcement provoked a rush of requests from other research establishments for taxol samples.

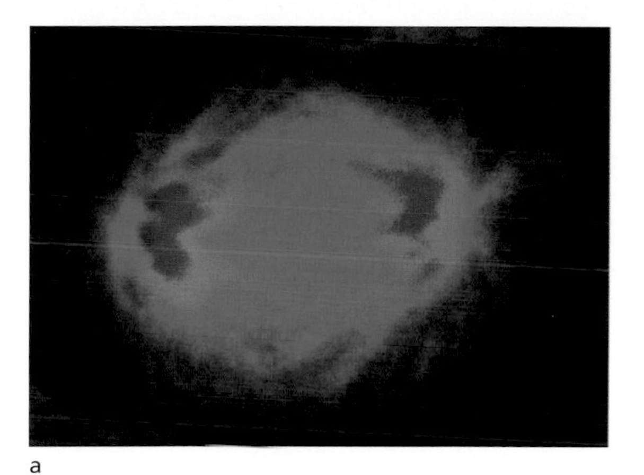

a

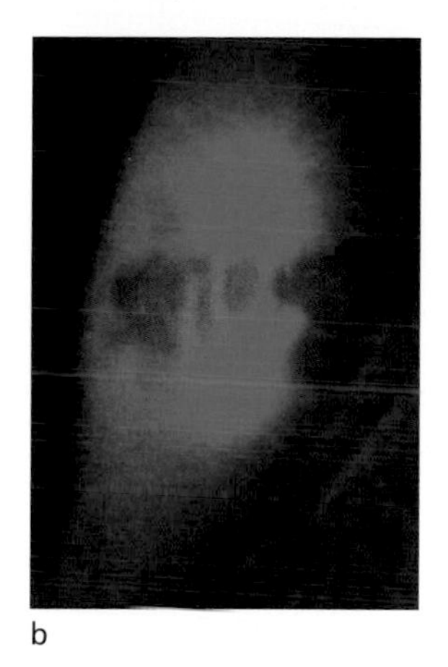

b

**Figure 7.9**

Before (a) a normal untreated cell divides, the genetic material (DNA, shown in blue) is duplicated and divided between the two halves of the cell, which subsequently splits into two. In (b) a taxol-treated cell, the mechanism that pulls the DNA to each end of the cell is disrupted, preventing proper cell division.

## Ecological plight of Pacific yew trees

The yield of taxol from Pacific yew bark is extremely low: 2,700 kg of dried bark yields only 3 g of taxol. As the various research trials proliferated, larger quantities of bark were required and the possibility of an effective anti-cancer drug meant that more and more bark would be required. Following World War II, there was a shortage of timber in the USA and ancient forests were opened up to logging. The large trees were removed and most of the rest of the vegetation, not considered to be of any commercial value, was burned. Pacific yew was among the 'valueless' group, so its felling for scientific research purposes was regarded as a useful by-product.

No serious environmental assessment had been made for Pacific yew so its numbers, habitat and range were not known. In 1977, 15,400 kg of bark were required for processing, causing the destruction of 2,000–3,000 yew trees. There were some signs of unease among the forest rangers and local environmentalists and some local yew tree preservation groups were set up. Ten years later, the demand had increased eight-fold to 122,000 kg of bark per year. Following successful clinical trials, in 1991 it was estimated that 100,000 trees needed to be felled to provide the raw material as taxol moved into commercial production. It was obvious that such a demand was unsustainable and that Pacific yew and its associated organisms in the forests of western USA were endangered. It was designated a threatened species and was afforded limited protection, but the collection of bark was still permitted. Discussions between conservationists, national wildlife departments and drug companies continued for several years until a solution presented itself in the form of the European yew *Taxus baccata*.

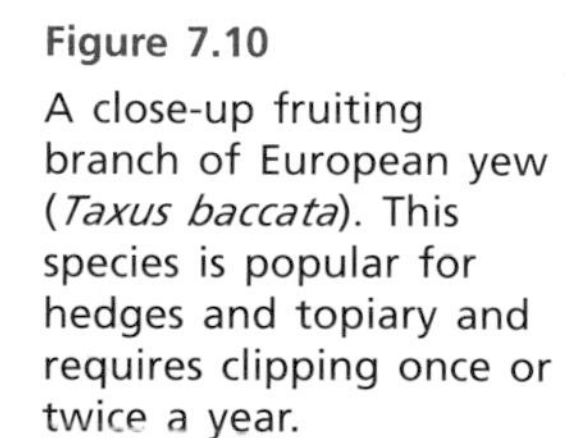

**Figure 7.10**

A close-up fruiting branch of European yew (*Taxus baccata*). This species is popular for hedges and topiary and requires clipping once or twice a year.

This European species has been known for centuries to be toxic to people and animals (Chapter 5). Screening of European yew found that a compound that could potentially be turned into taxol could be extracted in significant amounts from the leaves. As yew hedges are clipped annually, these clippings can be used without any damage to the trees. Several lines of research were undertaken to investigate if taxol could be produced from this material. Eventually, Robert Holton of Florida State University successfully developed a semi-synthetic process. In 1995, collection of bark from the Pacific yew ceased and the species is no longer under threat from unsustainable use. In the UK, several companies collect the clippings from parks and estates with large hedges of yew. The clippings must be from twigs under one year old as these leaves have the highest concentration of the compound that can be used to produce taxol.

## 7.5 The story of the African cherry

Ethnopharmacology is the study of traditional medicines and it can sometimes lead to the development of an effective drug treatment. In the past, there was an emphasis on finding plants that might be the source of new drugs for western medicine, with pharmaceutical companies and consumers/patients in the west being the main beneficiaries. Today, if a drug is developed in this way, the local communities should benefit, for example by a revenue-sharing agreement.

**Figure 7.11**

A felled African cherry tree (*Prunus africana*) in Cameroon with its bark stripped. African cherry trees are widely distributed throughout central and southern Africa and the islands around Madagascar at altitudes of 700–3,000 m. It is an evergreen hardwood tree that can grow to 25 m when mature. Flowering occurs between October and May, and the bitter cherry fruits are produced following insect pollination.

**Figure 7.12**

Medicinal plant stall in Malabo on the island of Bioko, Equatorial Guinea selling, among other things, African cherry bark.

### Traditional uses

Branches of African cherry (*Prunus africana*) are used for handles for tools such as axes and hoes, and felled trees are used in house building for flooring, furniture and other uses such as chopping blocks. Because the wood has a long straight grain, the timber is frequently used as the decking in bridges and trucks. The bark has been used in medicines for humans and for animals for many generations. African cherry has been an essential component of the traditional medicine chest in many parts of Africa for treatment of ailments ranging from fevers to chest pains and insanity. The leaves and fruits provide food for wildlife. The bitter cherries are not used by people but are eaten by mountain gorillas, monkeys, squirrels and birds.

## Modern use

About 45 years ago, an extract from the bark of African cherry was found to have a beneficial effect on benign prostatic hyperplasia (BPH), a non-cancerous enlargement of the prostate gland. This condition occurs in older men when the obstruction of the urethra impedes urination and emptying of the bladder. The bark extract has been found to contain several substances that are pharmacologically active and are known to have anti-inflammatory properties. No single active substance has been found to be effective against BPH. It seems that the compounds in the bark extract work together to control the condition by counteracting the biochemical and structural changes being produced in that region of the body. Clinical trials have shown that the bark extract is effective, well tolerated by patients and has very few side effects. It is used widely for the treatment of BPH in Europe. Attempts have been made to move from harvesting the bark unsustainably from wild African cherry trees to harvesting from cultivated trees or from wild plants using more sustainable methods. However, this has been met with mixed success.

## Effects of bark collection

In theory, bark may safely be removed from African cherry without harm to the trees, as long as the cambium layer, from which new bark grows, is left intact. But research in several countries, including Cameroon and Madagascar, has shown that even when harvested carefully, the growth of the trees is impaired. Without the protection of a thick layer of bark, larvae of wood-boring beetles can invade the tree, damaging the active xylem and phloem tissue and resulting in the crown (the top of the tree) dying back. Trees that have had bark harvested do not attain the height and girth that would be expected were they not stripped. Demand for the bark is very high and the prices paid for it are attractive to the people of the countries where it grows naturally. It has been estimated that 2,000 kg of fresh bark is needed to produce 5 kg of bark extract. This has resulted in over-harvesting and African cherry trees are now regarded as being vulnerable, just one step away from being classed as endangered. In some areas, plans have been drawn up to harvest the bark only from trees over a certain age and size, and only to strip the bark from patches on opposite sides of the trees. Where the local people have a sense of ownership, and an interest in the survival of the trees, this scheme has worked well. However, in many places, outsiders have come in and completely removed the whole bark from trees or have felled them in order to strip all the bark.

**Figure 7.13**
Illegally collected and subsequently confiscated African cherry bark in Bafia Town, Cameroon.

It is difficult to monitor trees scattered throughout a forest, and it may take a long time even to register that the bark 'pirates' have been at work. With 22 countries being home to African cherry there are many different management schemes in operation. In certain areas, benefits from African cherry commercialisation have translated into poverty alleviation at a community level through the creation of cooperatives that carefully manage how much bark is harvested, while at the same time achieving a reasonable price for it. Demand for the bark extract is starting to decrease, due to problems in producing extracts of consistent quality, but trees are still being exploited for their bark, albeit to a lesser extent.

### Plans for the future

Some studies have been carried out, particularly in Cameroon, and recommendations made to conserve the trees and control the trade of bark for medicinal purposes. The recommendations include:

- country-wide surveys to assess the numbers and health of populations of African cherry
- local conservation, with community management plans that limit harvesting to sustainable levels
- only allowing bark to be stripped from trees above a certain trunk diameter
- the establishment of plantations that can eventually be stripped in rotation, thereby providing a regular supply of bark and a steady income for the farmers; the trees can be grown close to other crops, except maize, without causing any problems
- the establishment of a monitoring service to ensure that sustainable practices are being used
- gaining a better understanding of the species' chemistry so that plants with the right chemical contents are propagated
- the possibility of using micropropagation (tissue culture) to produce large numbers of young trees quickly.

## 7.6 Sustainability and livelihoods

Mass harvesting of material from wild plants has nearly devastated the populations of species such as Pacific yew and African cherry, and these stories should elicit caution in the handling of future discoveries of potential medicinal plants. It is important that once a lead from a natural resource has been identified, supply issues are addressed and clinical trials undertaken.

As plant-based remedies are becoming more popular in some parts of the world, and there is less growing land available, the

**Figure 7.14**
*Hoodia gordonii* is one example of an over-harvested medicinal plant. Used for centuries by the indigenous peoples of the Kalahari Desert in southern Africa, it has gained wider popularity in the past 10 years as an appetite suppressant, and wild populations are now under threat from over-harvesting.

supply of good-quality plants can be an issue. A key area of concern is sustainable harvesting; in other words, maintaining harvesting at a level that allows continued collection in the future. If medicinal plants are over-harvested, species may face local extinction, and human communities that depend on the plants for their livelihoods suffer. In Nepal, harvesting of medicinal plants makes up 15–30% of the income of poorer families.

Ecological assessments of harvesting processes and yields can lead to community management plans. An alternative is cultivation of medicinal plants in gardens and fields. This has proved very effective in increasing yields, but the benefits do not always flow back to the original wild-harvesting community.

## 7.7 The globalisation of traditional Chinese medicine

The practice of traditional Chinese medicine (TCM) dates back over 2,000 years, its evolution being well documented in the vast numbers of local and national medicinal texts called 'Ben Cao' (herbals). These texts not only describe the underlying clinical theories such as 'yin' and 'yang', the five-elements theory, and how to formulate each multi-ingredient remedy; they also describe the species and plant parts used. An elaborate naming system has developed for each plant ingredient which, as well as describing its appearance, often reflects its geographical origin and any processing it may have undergone (for example stir-frying with honey or soaking in yellow rice wine). Such processing may be done for a number of reasons: to improve the efficacy of a herb, reduce its toxicity, or enhance the overall performance of a multi-ingredient formula by chemically interacting with other ingredients.

Under Chairman Mao Zedong, TCM was formally adopted by the Chinese government. In the early 1950s, TCM began to be standardised through the publication of a national pharmacopoeia, revised every five years. TCM continues to thrive in modern China with over 2,800 TCM hospitals, and TCM provides primary healthcare for around 70% of the population. Thanks to heavy investment from central government in recent years, this US $30 billion industry is now undergoing a new phase of rapid innovation and modernisation with the manufacture of TCM products expected to double within the next 10 years. At least 30% of current output is destined for overseas markets. Traditionally these have included other parts of South-East Asia such as Japan, Vietnam, Malaysia and Thailand, but since the early 1980s the market has expanded significantly to include many European countries, the USA, Canada and Australia where public demand for alternative and complementary medicines has spiralled. Today, TCM is practised in over 160 countries. In the UK alone, as many as 2,000 clinics can be found offering a wide range of treatments especially those for chronic conditions, such as eczema and psoriasis, for which western medicine has yet to provide long-term solutions.

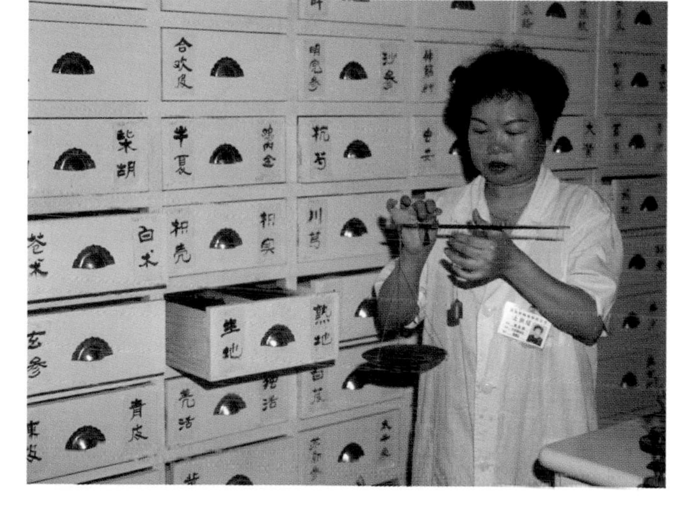


**Figure 7.15**
A Chinese apothecary preparing traditional Chinese medicines.

## The challenge of herbal authentication and the plant names maze

The rapid entry of TCM into new cultures around the world has brought with it both opportunities and challenges. The absence of statutory practitioner and product regulations in western countries, to ensure that TCM is practiced safely, is one of the biggest challenges. There is also an urgent need to develop herbal authentication and quality assurance systems and to standardise herbal ingredient names.

The traditional names of TCM herbs are often complex, and there is a further challenge in that a single TCM herb name may refer to more than one species with similar clinical profiles. For example, Bai Bu may officially be sourced from *Stemona sessilifolia, Stemona japonica* or *Stemona tuberosa*. Unofficial substitute herbs are also traditionally used in China if the official species are in short supply, or if there is local clinical preference for using another species. In China, such variations are the norm, and local, often highly trained, TCM doctors will prescribe accordingly. But this knowledge of herbal substitution and the complex plant naming system is difficult to transmit to foreign cultures and, given the absence of an international standardised TCM plant naming system, confusion over the identity of TCM ingredients will inevitably arise. The lack of government-recognised qualifications for herbal practitioners and herbal dispensaries in many western countries further complicates the situation and exposes the public to TCM medication that may contain incorrect and potentially harmful species.

A spate of serious adverse reactions occurred in the EU in the late 1990s involving TCM herbs in the genus *Aristolochia*, known in TCM as types of 'Mu Tong' or 'Fangji'. Amongst these cases were two patients in the UK who suffered kidney failure following the use of TCM formulae prescribed for eczema. Using the TCM reference collection at the Royal Botanic Gardens, Kew and the expertise of staff in natural product chemistry, Kew was able to identify the causal plant ingredient as *Aristolochia manshuriensis* (Guan Mu Tong) and to demonstrate the presence of the renal toxins, aristolochic acids 1 and 2, which this species is known to contain. Mis-identification at the point of prescribing and dispensing appeared to be the cause of both cases of adverse reaction. Also at fault was the absence of any mandatory regulations to ensure the competence of herbal practitioners and dispensary staff alike. Other serious cases involving related herbs in the *Aristolochia* group occurred elsewhere in the EU around the same time; in the UK this resulted in 1999 in the banning all 'Mu Tong' and 'Fangji' herbs, with China and other countries following suit soon after.

**Figure 7.16**

A range of plant products used within traditional Chinese medicine.

## Authentication progress in the west

One of the largest gaps in TCM quality control outside China has been the lack of easy access to authentic Chinese medicinal plants, notably in the form in which they are traditionally prepared as medicines, which is as dried and processed plant parts.

For 250 years, the Royal Botanic Gardens, Kew has developed plant collections from around the world to advance botanical knowledge, especially about useful plants, and to promote their conservation and sustainable use.

This provided the context for the Chinese Medicinal Plants Authentication and Conservation Centre (CMPACC), which was launched at Kew in 1998, in conjunction with its Beijing partner, the Institute of Medicinal Plant Development.

CMPACC's aim is to provide a scientific herbal reference and authentication resource for:

- protecting patients from receiving incorrect herbal ingredients
- developing standards for herbal identification, quality and labelling
- designing herbal quality assurance and drug monitoring (pharmacovigilance) systems
- natural product research
- training student practitioners and conservation groups
- promoting conservation, traceability and sustainable harvesting methods.

To date, CMPACC's extensive field programme to collect medicinal plants in China has resulted in a 'materia medica' resource of over 500 species, representing about 80% of the plants included in the *Chinese Pharmacopoeia*. All have been collected and processed according to TCM tradition and cross-referenced to dried herbarium specimens to provide proof of identity. In addition, first-hand investigation of TCM herbs sold in the markets and clinics, both in China and the UK, has led to the inclusion of unofficial substitute and counterfeit herbs, together with a better understanding of the complexity of plant names used by the TCM trade.

The persistent use of counterfeit herbs for certain species typically occurs when there are supply problems with the genuine article, perhaps due to over-harvesting of wild populations. One example is the heartwood of the tropical eaglewood tree *Aquilaria sinensis* 'Chen Xiang', used in TCM for stomach complaints. It is now so scarce in the wild in southern China that counterfeits made from black-painted wood of unknown origin are found in street markets and clinics; the black paint is an attempt to mimic the black resin-filled wood so typical of genuine Chen Xiang.

**Figure 7.17**
Anguo medicine market in Hebei Province, China.

Including market samples, CMPACC's reference collection currently totals over 3,500 items. It also demonstrates a range of processing types for a single plant part, such as slicing, carbonising, or cooking with ginger, bran or honey. To complement the use of the collection for authentication using morphological characters, each sample has also been chemically fingerprinted and DNA extracted by Kew's on-site laboratory. This provides 'fingerprint libraries' which are especially useful for examining ingredients in powdered, granule or pill form and in multi-ingredient preparations. New DNA fingerprinting methods (similar to those described in Chapter 5) have the potential to offer powerful authentication markers as well as tools for detecting herbal contaminants. As interest in TCM herbs continues to grow, so too does the source of authentication enquiries; to date these range from medical practitioners, health regulators, academics, conservation groups as well as the medical, food and cosmetic industries.

As an authority on plant names, Kew is also establishing a Medicinal Plant Names Index and Information Service (MPNI) that will bring together the miscellany of plant names that can be used to identify a herbal ingredient.

# 8 Drink and drugs

The key role that plants play in our nutrition has already been covered. Plants are also the source of many drinks used as stimulants, rather than simply for nutrition. This chapter describes some of these drinks and investigates the history of the use and cultivation of the plants that gave rise to these drink products. In the cases outlined, the extreme popularity of the resulting plant products has promoted the global movement and cultivation of the relevant plant species. Plants can also produce chemical compounds that are used by humans for their ability to affect mood and emotional state, so called psychoactive properties. The example of cannabis is considered, together with the physical and mental responses induced.

## 8.1 Tea

Every day, people around the world consume over 3 billion cups or glasses of tea, more than any other drink except plain water. Until the sixteenth century, tea was scarcely known outside eastern Asia. In 1664, the first significant consignment of tea reached England from China under the auspices of the East India Company. But when this company lost its monopoly on trade with China, it sought to establish an alternative source of the crop in India where tea cultivation began in the 1830s. Tea was introduced to East Africa in the 1880s.

Two varieties of tea plant produce most of the world's tea: *Camellia sinensis* var. *sinensis* and *Camellia sinensis* var. *assamica* (var. = variety). *Camellia sinensis* var. *sinensis* originated in China and was the first tea plant to be cultivated. This variety grows as a bush or dwarf tree that has small leaves. *Camellia sinensis* var. *assamica* is native to Assam, Burma, Thailand and Vietnam. It is faster growing and has larger leaves than var. *sinensis*, and became the most widely cultivated variety in India. Both varieties need tropical or sub-tropical climates with plenty of rainfall but cannot tolerate frost. Tea is grown in more than 30 countries, with most still produced in China and India, which are also the largest consumers. Kenya is now the world's largest net tea exporter.

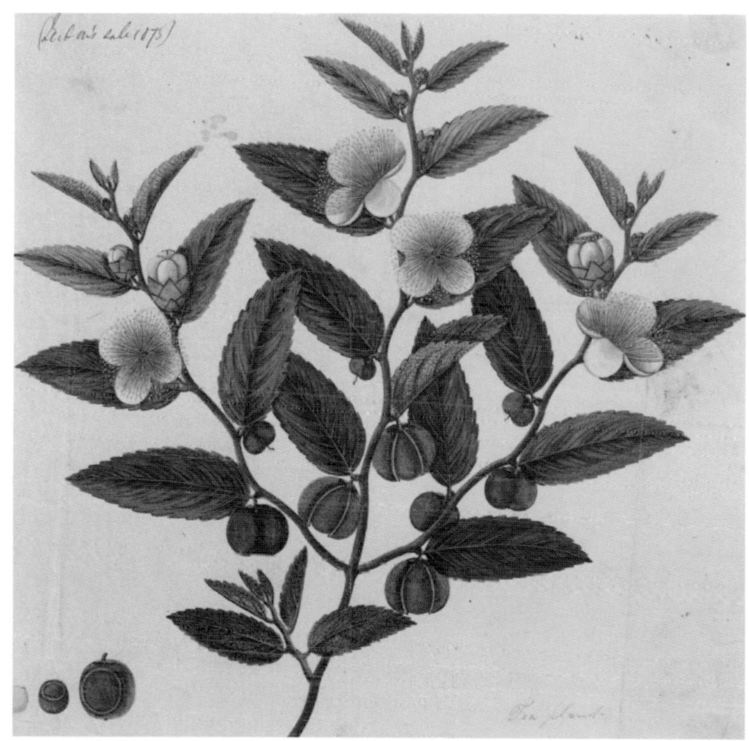

**Figure 8.1**

Tea leaves, flowers and fruit. Watercolour on paper, artist unknown, c.1840.

**Figure 8.2**

Tea being harvested in a plantation. Tea is harvested every week or fortnight throughout the year, with pickers plucking by hand the shoot tips, comprising just the two top leaves and terminal bud. A skilled tea picker can harvest up to 35 kg of shoot tips per day, which is enough to produce about 9 kg of dried tea. Global production of tea in 2007 was 3.9 million tonnes.

Different types of tea (black, green, white or oolong) are produced by processing the harvested leaves in different ways. Most of the world's tea production is in the form of black tea. The leaves are withered in warm air and then broken up by machine to release oxidising enzymes that speed up reactions between chemicals in the tea leaves and oxygen in the air. During these reactions, which are known as 'oxidative fermentation' (although they differ from the fermentation process that results in alcoholic drinks), the leaves change colour to golden-red over a period of 3–4 hours. After drying to a characteristic brown or black, the leaf particles are graded by size.

By contrast, green tea is not fermented. After withering, the leaves are heated or steamed to prevent oxidative fermentation and then rolled and dried. These leaves produce a pale coloured drink with a mild taste and delicate flavour. White tea is made in China from very young leaves and buds that are simply withered and then dried. Brewed white tea is pale yellow in colour and has a subtle flavour. Oolong tea undergoes a shorter period of fermentation than black tea and produces a delicately flavoured drink.

To make a drink of tea, in many countries, the leaves are simply infused in boiling water. Tibetan nomads use chunks of 'brick tea', ground tea which has been steamed and pressed into shape; they boil this for several hours producing a thick black brew, to which salt or soda is added, with a final flavouring of rancid yak butter. Other flavourings added to tea include oil of bergamot orange (*Citrus* x *limon*), as used in Earl Grey tea; jasmine flowers, which are added to oolong tea; and peppermint leaves, which are added to green tea. In India and East Africa, many people drink 'chai', which is made by boiling black tea, water, sugar and milk together with spices such as cinnamon, cloves, ginger or black pepper.

**Figure 8.3**

Black, green and white teas.

## The chemistry of tea

Tea's stimulant effects come from caffeine and a compound called theophylline, which are both found in the leaves. Theophylline has been used medicinally to treat asthma, as it acts as a bronchodilator, relaxing airway walls and relieving coughing and wheezing. Tea leaves also contain high levels of flavonoids, some of the antioxidants that were described in Chapter 6, and a series of flavanols known as catechins. When black tea is fermented, catechins are converted to polyphenols, which contribute to the colour of the tea. These compounds are water-soluble, so longer brewing during tea-making will give rise to higher concentrations of these compounds in the liquid. Other constituents of tea leaves include tannins, which are responsible for the dark stains caused by tea and for its astringent taste.

## 8.2 Coffee

Reputedly, the first person to experience coffee's stimulant effect was an Ethiopian goat-herd, who witnessed his goats becoming increasingly lively as they browsed on the fruits of a small shrub and then decided to try out the fruits himself. This shrub is what we now know as coffee. At first, people chewed the fruits or crushed them with fat as a food, or made a drink from the leaves. Records dating back to the thirteenth century have been found in Yemen and these mention the roasting of coffee seeds and the grinding of them to make a drink. Religious worshippers drank coffee to keep themselves awake during lengthy ceremonies.

Until about 1600, coffee was cultivated only in North Africa and the Arabian Peninsula, but it was then introduced to India. It was another 100 years before coffee plants were exported to Java. Six years later, the botanic garden in Amsterdam received a single plant from Java. The seed it produced was sent out to the Dutch colony of Surinam in the Caribbean. The French also received seeds from the Amsterdam plant, which they grew on and, in turn, they sent plants to Haiti and their other West Indian colonies.

Brazil, now the world's main coffee producer, began growing coffee in 1727, also using plants grown from the seeds from the Amsterdam plant. The narrow genetic base of coffee growing in the Americas, due to the cultivated crop resulting from the introduction of a single plant, became a major issue with the onset of coffee leaf rust disease. All of the coffee cultivars grown in Latin America showed little or no resistance to the disease. Breeding programmes are now identifying cultivars that have some resistance to coffee leaf rust, and introductions of plants from outside of the Americas are being sought in order to widen the genetic base.

The best quality coffee is produced by *Coffea arabica*, which is native to south-western Ethiopia and parts of Sudan. The West African species *Coffea canephora* produces robusta coffee, which has a stronger flavour than coffee made from *Coffea arabica*. Robusta coffee is also resistant to coffee leaf rust. Coffee breeders are using the 'hybrid de Timor', a cross between arabica and robusta, to produce cultivars combining the quality of arabica with the disease resistance of robusta. Coffee will only grow in tropical areas with rainfall between 150 and 250 cm per year. *Coffea arabica* thrives in the cooler highlands, whilst *Coffea canephora* can withstand hotter drier conditions.

**Figure 8.4**
Coffee is a small evergreen tree or shrub, bearing fragrant white flowers. Its fleshy fruits, which turn red as they mature, generally contain two seeds known as 'beans'.

**Figure 8.5**

Before they are roasted, green coffee beans have very little flavour or scent. Roasting beans at different temperatures and for varying lengths of time produces flavours and aromas that appeal to individual tastes. Generally, green coffee beans are heated to 180–240 °C for 8–15 minutes. As the roasting progresses the beans become darker, and once they reach a specific moisture content they 'pop'. Roasting converts starch to sugar, breaks down proteins and releases the caffeol (coffee oil), which produces the flavour and aroma. The world's major producers are Brazil, Colombia, Vietnam and Indonesia. About 8 million tonnes of coffee are produced worldwide each year, and every day about 1.5 billion cups of coffee are consumed.

After picking, the coffee fruits are treated to separate the pulp from the seeds (beans). Two different methods are used, depending on the availability of water for processing. The resulting green coffee beans are dried, and then cleaned to remove any contaminating material before being prepared for export. The beans must also be roasted before they can be used.

The amount of caffeine in a cup of coffee depends on the origin of the beans, the method of brewing and the strength of the brew. Robusta coffees contain approximately twice the amount of caffeine as arabica coffees. A cup of roast and ground coffee brewed by the filter method contains on average 115 mg caffeine. Instant coffee generally contains about 65–100 mg caffeine in a cup of the same size. Tea is higher in caffeine by weight than coffee but less tea is generally used to make a cup of drink so the caffeine concentration is lower. Cola drinks generally contain about 120 mg of the stimulant per litre.

## Caffeine and its effects

Caffeine is a plant alkaloid and a psychoactive drug that is present in many plants. Caffeine also acts as a natural pesticide: many insects are deterred from feeding on plants containing this compound and other insects that feed on plant tissue containing caffeine become paralysed.

Consumption of caffeine by humans can cause a temporary increase in levels of alertness and ability to concentrate by acting on the central nervous system, also causing the release of the 'fight or flight' hormone adrenalin. Many of the drinks preferred by humans worldwide contain caffeine, for example tea, coffee and cola. It is possible to remove the caffeine from drinks to produce decaffeinated versions that still contain the other flavours, and people who find that caffeine keeps them awake at night sometimes prefer these. Decaffeinated coffee should contain less than 0.1% caffeine by weight in the roasted beans. Caffeine is extracted using various solvents, including water. In water-decaffeination, green coffee beans are soaked in water containing all the water-soluble components of coffee except caffeine, so that these compounds are not removed from the beans along with the caffeine. Caffeine diffuses out from the beans into the water and can be removed from the solution using activated charcoal. Tea can be decaffeinated by a similar process.

## 8.3 Cola

Cola drinks are the world's most popular soft drinks. In 2008 the average US consumption of soft drinks was 216 litres per person: cola made up one-third of this volume, while fruit juices made up just one-quarter. Cola-flavoured drinks first became popular in the late nineteenth century when a pharmacist combined extracts from kola nuts and coca leaves (*Erythroxylum coca*) with sugar, carbonated water and other flavourings to create a 'brain tonic'. Today, most cola flavourings are artificial, although a few brands still contain extracts of kola nuts. Coca leaves are rarely used as they have to be treated to remove any trace of cocaine or the chemicals used by the plant to synthesise cocaine.

**Figure 8.6**

Seed pod of *Cola nitida* encasing several seeds. Kola nuts are the fruit of a genus of trees (*Cola*) found in the tropical West African rainforests. The genus belongs to the same family as cocoa, the Malvaceae family. Two species are widely cultivated for their seeds, *Cola acuminata* and *Cola nitida*. The trees, which reach 15 m in height, bear clusters of yellow flowers followed by large multi-seeded pods. After the pods are harvested, the seeds are fermented so that the seed coat breaks down to reveal the cotyledons. Within the trade, these are known as kola nuts. Most kola nut cultivation still takes place in West Africa, but nearly all of the nuts are exported to other sub-Saharan countries. Some kola nuts are exported to Europe and North America, mainly for the production of cola flavourings.

The active constituents of kola nuts are the alkaloids caffeine and theobromine, and various flavonoids. After chewing the nuts, people find that water tastes sweet, a useful property for those living in areas with poor water quality. Chewing the nuts also helps to reduce hunger and suppress appetite, but frequent chewing of kola stains the mouth and teeth red. Kola twigs are used as chewing sticks to clean the teeth and freshen the mouth. Bark of both *Cola acuminata* and *Cola nitida* has been used to treat dysentery and chest complaints, and bark extracts have been found to have a beneficial anti-bacterial activity.

### Fairtrade of tea, coffee and cocoa

The concept of Fairtrade has developed globally over the past 30 years with a network of organisations around the world working towards 'better prices, decent working conditions, local sustainability, and fair terms of trade for farmers and workers in the developing world' (UK's Fairtrade Foundation's website). Virtually all tea, coffee and cocoa production occurs within countries that can be classed as developing. Fairtrade seeks to ensure that producers in these countries benefit from increased direct access to markets and receive a guaranteed minimum price that covers the cost of sustainable production, enabling community projects such as schools and health centres to be funded. Goods sold in the developed world under the Fairtrade scheme are labelled as such. In the UK, during the early 1990s, brands of coffee, chocolate and tea were the first products to be certified as being Fairtrade. The worldwide market for such goods is growing quickly; in 2008 it was around US $4 billion.

## 8.4 Cocoa: 'food of the gods'

Cocoa is one example of a plant product that has been used to generate both a drink and a food, chocolate. A drink of roasted cocoa seeds (beans) mixed with ground maize and chilli peppers may not sound too appetising at first. But the emperors, nobility, warriors and priests of the Aztec peoples of Central America prized this concoction, called chilcacahuatl. They considered the cacao tree to be a gift from the gods and valued its seeds very highly. The Aztecs even used the seeds as a form of currency: 30 cocoa beans would buy a small rabbit.

**Figure 8.7**
Raw, dried cocoa beans before processing.

In the middle of the sixteenth century, Spanish soldiers and navigators transported cocoa beans to Europe. The Spanish also made a drink from the beans, but they flavoured it with sugar and vanilla. By 1650, hot drinking chocolate was being served throughout Europe. When the Swedish botanist Linnaeus gave the tree its scientific name, he chose one that reflected the traditional beliefs of the Mayans, calling it *Theobroma cacao*; the genus name *Theobroma* is derived from the Greek words meaning 'food of the gods'.

The cacao tree is native to the upper Amazon basin, in the foothills of the Andes, but its first centre of cultivation was in Central America. It requires a hot tropical climate to thrive, and today most of the world's large-scale cocoa production takes place in West Africa and South America.

When first harvested, the bitter-tasting beans do not have the flavour or aroma of chocolate. They are heaped up in the sun and allowed to ferment for up to a week, during which time the characteristic flavour and colour of the beans begin to develop. This process is similar to the oxidative fermentation to which tea and coffee are subjected. After drying, the cocoa beans are usually transported away from the farm or plantation to the chocolate factory where they are roasted to achieve the deep brown colour, flavour and scent that are so enjoyed by chocolate-lovers around the world. There are some 550 chemicals responsible for flavour in chocolate.

Further processing is required to produce the chocolate available in shops. When the roasted beans have been cracked open to release the two cotyledons (known as nibs), these are crushed to a paste and some of the oily cocoa butter is extracted for use in white chocolate. It is also used in cosmetics and toiletries.

The solid cocoa mass that remains after cocoa butter has been extracted is now ready to be used in baking or making sweets. It can be dried, ground down and used to form cocoa powder. However, much of the cocoa mass goes to produce chocolate. Dark chocolate consists of cocoa mass plus cocoa butter and sugar; milk chocolate contains these basic ingredients plus milk solids and other vegetable fats. At this stage, the chocolate is still rather gritty; a process called conching, in which melted chocolate is slowly stirred and kneaded, produces the smooth silky texture of the finished product.

**Figure 8.8**
The cacao tree (*Theobroma cacao*) is a small tree, generally no more than 8 metres high, which can grow in the shade of other taller trees. Its tiny white flowers appear directly on the main trunk and older branches and are pollinated by midges. The rugby-ball shaped fruits develop over 4–6 months and turn yellow or orange in colour as they mature. Inside the fruit's leathery skin is a creamy sticky pulp enveloping the ivory coloured seeds, commonly known as beans.

Conching also helps to develop the flavour. Finally, the chocolate is heated to a specific temperature and cooled slowly (tempered) so that it becomes hard and glossy.

Since the seventeenth century, chocolate and cocoa have been reputed to have an enormous range of medicinal properties. Some of the earliest British chocolate makers (Fry of Bristol and Terry of York) were apothecaries who became interested in using chocolate to treat various conditions. Like red wine and tea, dark chocolate contains beneficial flavanols called catechins.

## Why does eating chocolate make you feel good?

Eating chocolate induces a feeling of well-being in many people, so much so that some people admit to being 'chocoholics' who crave it. Among the many qualities attributed to chocolate are its activities as a stimulant, relaxant, aphrodisiac, tonic and antidepressant. Its stimulating qualities are due to caffeine and theobromine. One of the most interesting constituents of chocolate is a chemical called phenylethylamine. Normally, phenylethylamine is produced in the brain and levels are known to increase when people fall in love. Phenylethylamine is also thought to have a role in controlling both appetite and mood in humans.

## 8.5 Beer

The drinks covered so far have not included any alcoholic ones and to many people beer, wine and the various spirits that are derived from plant sources are as important as the non-alcoholic ones. Alcoholic drinks are made by utilising the sugars or starches present in plants and subjecting them to a process known as alcoholic fermentation, a different process from the oxidative fermentation used in the production of tea, coffee and cocoa.

### Alcoholic fermentation

All alcoholic drinks arise from plant material rich in either sugar or starch. If the material is starch, such as that in the barley grains used to produce beer or whisky, then the starch needs to be converted to sugar before fermentation can occur. For fruit-based alcoholic drinks, such as wine, the fruit itself already contains sugars. Alcoholic fermentation relies on yeast converting the sugar into alcohol. The yeast is either added directly, as in brewing of beer, or, in the case of wine, is already present on the skins of the grapes.

Beer is perhaps the oldest alcoholic drink, being referred to 5,000 years ago in writings on Sumerian clay tablets. Beer is relatively simple to produce; it requires only malted cereal grains (although other sources of starch can be used), water and yeast. In the malting process, the cereal grains are allowed to germinate, which causes the stored starches they contain to be converted to sugars by enzymes. The resulting malted grains are then dried prior to being treated with hot water to allow any further sugar and flavour components to seep out. The sugary liquid is then boiled, both to reduce its volume and also to destroy any remaining enzymes. Hops (*Humulus lupulus*) can be added to give flavour and characteristic bitterness before the mixture is cooled and yeast is added to allow fermentation to occur and alcohol to be produced.

**Figure 8.9**
Hops (*Humulus lupulus*) are used to give flavour and characteristic bitterness to beer.

Alcohol is a natural preservative which prevents many bacteria and fungi from growing. Nevertheless any alcoholic drink, if exposed to air, will oxidise and this affects its taste. Before treatment of drinking water became the accepted norm, it was safer for the population, including children, to drink weak beer rather than water. Beer is the third most popular drink worldwide, behind tea and coffee. In 2006 global beer production was 133 billion litres.

## The effect of alcohol on the human body

Alcohol, or specifically ethanol, is very easily absorbed by the body once consumed. It travels in the bloodstream to the brain where it has a number of effects. At low levels it causes people to feel relaxed and happy by stimulating those areas of the brain responsible for controlling mood. At higher doses the effects can be different. The areas of the brain responsible for muscle movement and learning become less responsive. The production of the anti-diuretic hormone responsible for controlling the amounts of water present in the body is also affected, which typically leads to excessive urine production and a feeling of dehydration. Over-consumption of alcohol on a frequent basis can lead to addiction, or alcoholism, whereby the body craves alcohol.

## 8.6 Cannabis

Chapter 7 covered how plants have had a long-established role in medicine. In addition to the pharmaceutical properties of many plant compounds, humans sometimes use plants for their psychoactive properties. Cannabis is perhaps the best-known plant that contains such compounds.

### History

The cannabis plant, hemp, has a long historical association with human civilisation in several ways. Hemp has been used to produce fibre for matting and ropes for many thousands of years. A mummified body discovered in the Xinjiang area of China and dated as 2,500 years old was found with a leather basket and a wooden bowl containing cannabis leaves and seeds. Other items found in the tomb suggest that the reason for the presence of cannabis leaves and seeds was that the body was of someone who had been a shaman, thought to be capable of communicating with the world of spirits, and aware of the psychoactive properties of cannabis. One of the alternative names for cannabis is ganga, which derives from the Sanskrit ganjika. This indicates early use of cannabis by the Hindus of India and Nepal.

### What is cannabis?

Two different products can be made from the cannabis plant and both have psychoactive properties:

- cannabis herb, or marijuana, consisting of the leaves or flowers that can be dried and then smoked
- cannabis resin, sometimes called hashish, which is the dried exudates, or secreted fluid, from small hairs associated with the flowers.

By far the most widely produced of the two is cannabis herb, which accounts for around 80% of the product seized globally by law enforcement officers.

Cannabis belongs to the Cannabaceae family. *Cannabis sativa* is divided into two subspecies (subsp.), *sativa* and *indica*. Recent crossing of these subspecies has led to the development of 'skunk', an extremely potent form of cannabis. Skunk plants have inherited advantages from both parents: high levels of psychoactive chemicals from *sativa*, and a rapid life cycle and high yield from *indica*.

a

b

**Figure 8.10**

(a) *Cannabis sativa* subsp. *sativa*.

(b) *Cannabis sativa* subsp. *indica*.

These two subspecies are quite different in appearance but both produce a number of psychoactive chemicals referred to as cannabinoids. Of the 70 cannabinoids uniquely present in cannabis, one known as delta-9-tetrahydrocannabinol, or THC, is responsible for the majority of the effects of cannabis in the human body. Cannabis plants are either male or female and it is the unfertilised female flowers that contain the largest amounts of THC. The unfertilised female flower buds are frequently harvested and described as 'sinsemilla' cannabis, from the Spanish for 'without seeds'.

## Facts and figures

Cannabis is considered to be the most commonly used illicit drug in the world. As both the growing and trading of herbal cannabis is illegal in many countries, it is difficult to obtain very accurate figures. However, the United Nations estimate that in 2004, 162 million people worldwide (4% of the global population) used cannabis. It is believed to be grown in 176 countries and it is estimated that some 45,000 tonnes was harvested in 2004. Estimates of the potential value of the North American market alone are US $10–60 billion for the 14,000 tonnes believed to be produced there.

Why is cannabis so widely grown and used? As a plant it can be easily grown in many conditions, is very productive and yields material that is in a form that can be used without further processing. As a result many cannabis users grow their own supplies.

**Figure 8.11**

Cannabis plants being grown indoors under powerful lights. Small-scale producers have been arrested after infrared night sights used on police helicopters have detected the large amounts of heat being emitted by the artificial lighting sources.

## The effect of cannabis on the human body

Although cannabis can be eaten to produce its effects, most users choose to smoke it. The high temperatures during smoking convert one of the cannabinoids into THC, which is then vaporised and absorbed more easily into the bloodstream. The THC in cannabis is fat-soluble and so passes quickly out of the bloodstream and into organs such as the brain. Here it acts on receptors and thus causes a number of physical effects on the body, such as increased heart rate, perceived changes in temperature of the hands or feet and a drying of the mouth.

One of the reasons that people use cannabis is because it can create feelings of euphoria, which then turn into a feeling of calmness, the so-called 'being stoned' effects. Other effects include heightened senses and an increased desire to eat. Negative effects for some people can include a sense of panic or confusion resulting from the changes that are occurring to them.

Under controlled laboratory conditions, cannabis has been shown to slow reaction time, make people more uncoordinated, reduce memory performance and increase the time taken to solve problems.

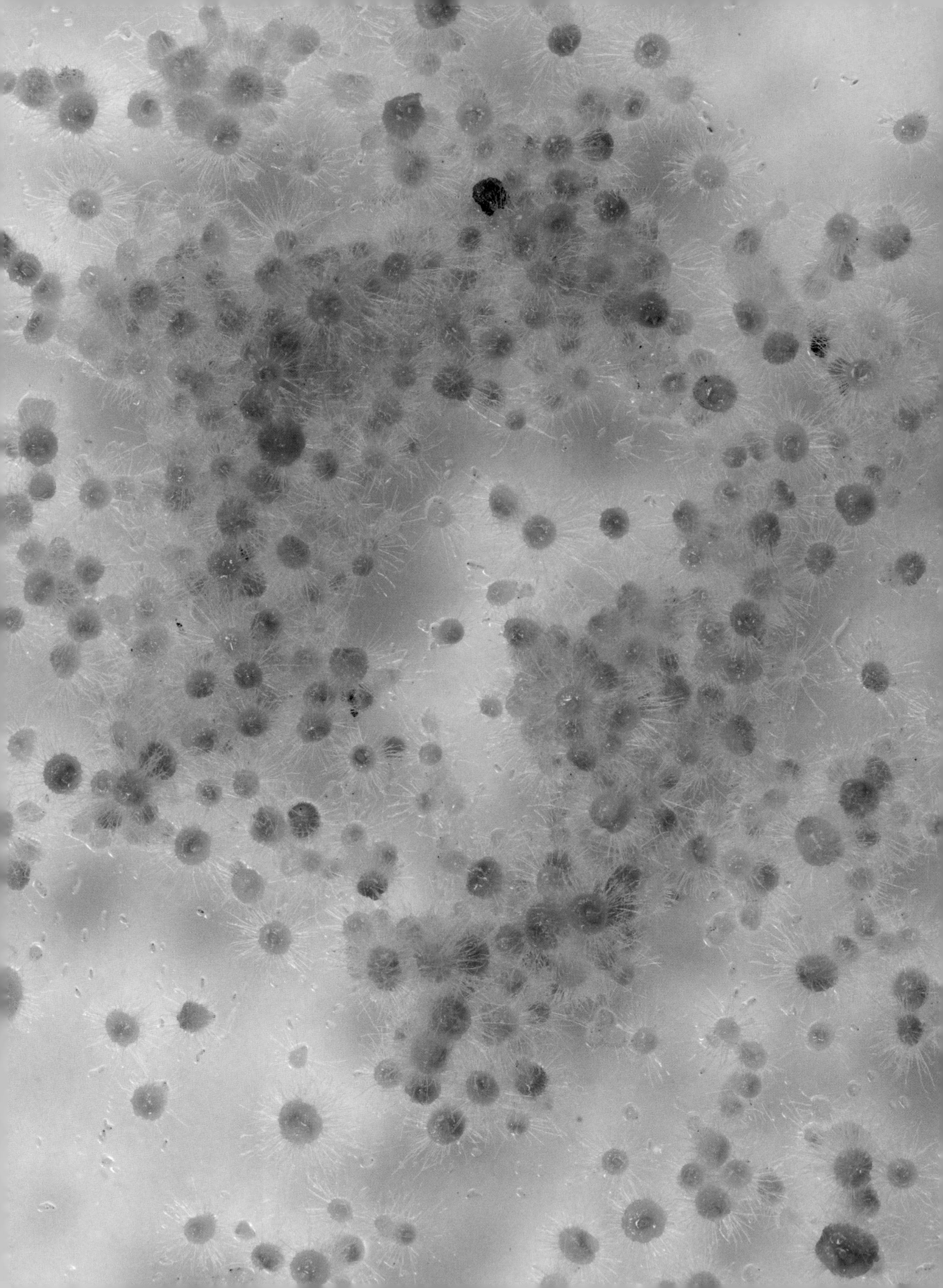

# 9 Micropropagation

Since about the sixteenth century, humans have been exploiting the natural ability of plants to reproduce vegetatively thereby generating new plants from cuttings. Taking, or 'striking', a cutting involves the removal of a small piece of the parent plant (leaf, stem or root) and placing it in an appropriate medium or soil. If the conditions are correct, the cutting will grow into a new plant genetically identical to the original parent plant. This is a form of artificial asexual reproduction and the progeny plants are 'clones' of the parent plant. Many horticultural techniques rely on this ability of plants to form whole new plants from small parts of an original plant, so-called totipotency. Scientists have taken this 'regenerative' property of plants one stage further in micropropagation (literally 'growing in miniature'), which exploits totipotency on a smaller scale and in a sterile container. This chapter discusses what micropropagation is and the ways in which the techniques have been successfully used for orchids and palms, in particular.

**Figure 9.1**

Totipotency is a naturally occurring phenomenon that can be demonstrated when a plant cutting is taken and new roots develop from the cut end of the stem, or when the buds on a horizontal strawberry (*Fragaria* hybrids) stem develop roots which subsequently form small plants around the original plant.

## How micropropagation works

Micropropagation describes growing plants on sterile media and can be used to germinate seeds or for tissue culture. This is where cells of different plant parts are stimulated to multiply and, as they each have the potential to form all cell types, they can produce roots, stems, flowers and so on. The starting point for tissue culture is known as an 'explant' and may be a leaf, bud or any other living part of the plant. Even a single plant cell can be isolated and cultured to generate an exact genetic replica or 'clone' of its mother plant.

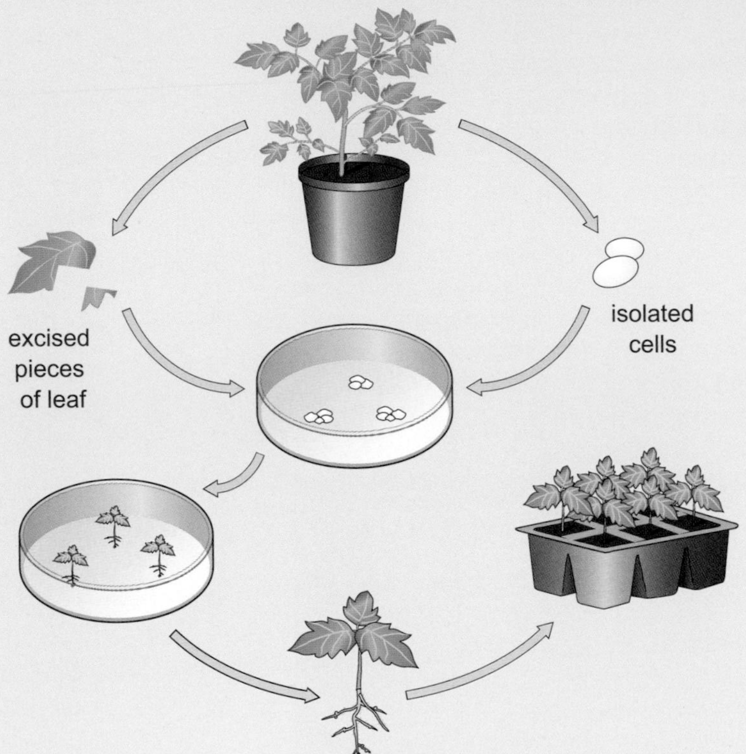


Figure 9.2
The basic process of micropropagation through tissue culture, using either isolated plant cells or pieces of leaf.

The explant has to be supplied with (cultured in) a growth medium containing nutrients and chemicals including growth regulators, which can control whether or not the explant produces roots or shoots. The tissue culture is carried out *in vitro*, meaning literally 'in glass', although these days is as likely to be in a disposable plastic container known as a petri dish. The petri dish contains agar, a jelly-like substance that acts as an inert medium to which plant growth regulators (PGRs) are added. PGRs are chemicals (they can be either natural or synthetic), which when applied to plants can have great effects on plant growth and development even at very low concentrations. Adding the auxin class of PGRs tends to result in more shoots being formed, whilst addition of the cytokinin class of PGRs results in more roots being formed. As an explant cannot photosynthesise sufficiently to support the regeneration of a new plant, sugar (typically sucrose) is added to the growth medium. Simple sugars, such as glucose and fructose, are not used as they can break down resulting in the formation of toxic compounds. Bacteria and fungi thrive on media containing sucrose, so tissue culture is carried out using sterile (aseptic) techniques to prevent the growth of these micro-organisms. As the plantlets grow they are transferred into larger containers with slightly different nutrients and plant growth regulators until they are ready to face life in the outside world.

a

b

Figure 9.3
(a) A micropropagation laboratory showing the plants being cultured in small plastic containers.
(b) A close-up showing plants growing within a sterile container.

## 9.1 Uses of micropropagation

Micropropagation is an important technique available to plant scientists, and its application can be demonstrated in several areas.

- In conservation, micropropagation techniques can be applied to culture critically endangered species and to assist in re-stocking habitats whose biodiversity is declining.
- In horticulture, the mass production of high-value plants, such as ornamental ferns, palms, and orchids, makes specimens available to the general public and reduces pressure on the remaining wild plants.
- In agriculture, the technique is widely used to propagate successfully large numbers of plants with desirable characteristics of yield, flavour, disease-resistance or drought-resistance for commercial plantations of crops such as potato and banana.
- Micropropagation is also used when plant cells have been genetically modified, in order to grow on plants that can then produce seed.

a

b

Figure 9.4

Two plant genera, *Drosera* (a) and *Nepenthes* (b), which have benefitted from micropropagation techniques.

Micropropagation allows rapid multiplication under controlled, pathogen-free conditions. It enables many plants that are difficult to propagate using conventional techniques, or whose seeds have complex dormancy requirements, to be produced in large quantities. Seed banks collect and store the seeds of low population plant species. However, there are some 'recalcitrant' plant species whose seeds cannot survive the process of drying and freezing before storage in the seed bank. One method of conserving these species is to remove the plant embryos from the seeds and store them directly in liquid nitrogen at –196 °C or to allow the embryos to grow and develop shoots and store these shoots in liquid nitrogen (so-called cryopreservation). If and when the embryos or the shoot tissue derived from them are needed, they can be taken out of the liquid nitrogen, thawed and grown in appropriate culture media by micropropagation.

Micropropagation techniques, growing in culture medium under sterile conditions, can also be used for very small seeds, such as orchid seeds, which are difficult to handle and germinate under standard conditions. For endangered species, every single seed is precious and micropropagation minimises seed loss. Carnivorous plants, such as sundews (*Drosera*) and pitcher plants (*Nepenthes*), have a low percentage of seed germination and the seedlings are very susceptible to fungal disease. The aseptic conditions and special care provided by micropropagation leads to more successful germination and greater survival rates.

**Figure 9.5**
*Cylindrocline lorencei* is a shrub that is restricted to Mauritius. Seeds were collected from the last remaining plants by staff from the Brest Botanic Gardens, France. Embryos were harvested from these seeds and plants were raised at Brest. By 1990, the species was extinct in the wild and existed only at Brest. In 2004, new plants were propagated at Kew *in vitro* by isolating dormant tiny buds from the plants received from Brest. In 2007, one-year-old plants were repatriated to Mauritius in the hope that they could be reintroduced into the wild.

## 9.2 Micropropagation of orchids

The Orchidaceae is one of the largest families of flowering plants and orchids are found in almost all countries. Many species have highly specialised adaptations to the particular habitats in which they live and are unable to survive elsewhere if that habitat is destroyed. Orchids have long been admired and collected as horticultural specimens but many are now under threat of extinction, mainly from over-collection of wild specimens but also because of habitat loss.

Most trade in wild orchids is prohibited for commercial purposes. Trade in artificially propagated plants is only allowed under very strict conditions and is subject to permit (this is covered in more detail in Chapter 13). Hence, micropropagation of orchids serves three purposes:

- as a conservation measure for plants whose habitat is under threat
- to provide material for research so that no removal of wild specimens is needed
- to provide plants for the horticultural trade.

Many orchids have complex interactions with insects and other animals which act as pollinators. In some cases, the orchid provides a source of food or shelter for insects. Most orchid species form relationships with fungi, which enable their tiny seeds to germinate. Orchid seeds are the smallest in the world so have insufficient food reserves to enable the embryo to grow and produce leaves. The seeds are dependent on nutrients from the closely associated fungi (known as mycorrhizal fungi) and, in return, once the leaves of the orchid have developed, the plant gives nutrients to the fungus. A two-way relationship like this is known as symbiosis.

There are two different methods of germinating orchid seeds in the laboratory: symbiotic and asymbiotic.

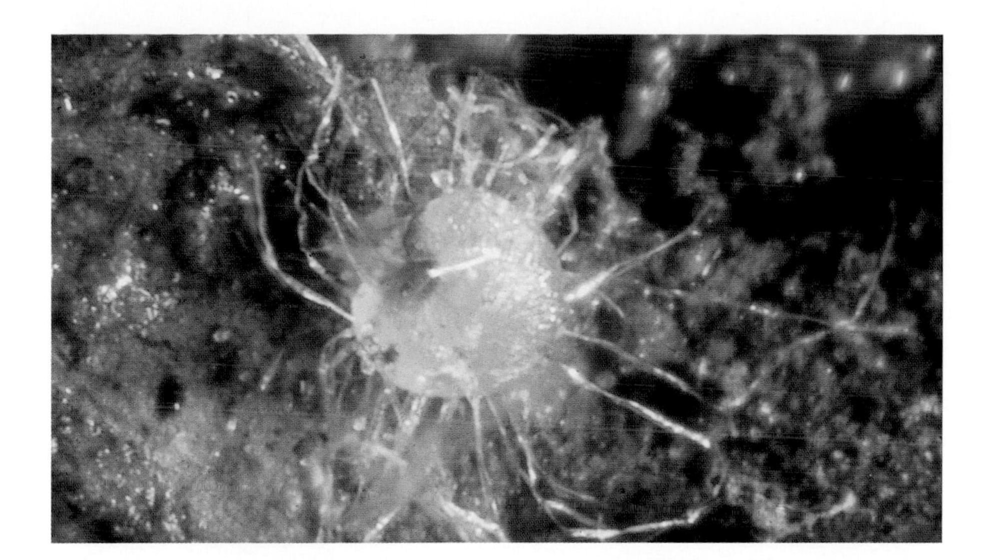

**Figure 9.6**

Germination of orchid seed in the soil is facilitated by symbiotic fungi. For micropropagation, these fungi are extracted from orchid roots and added to the sterile nutrient medium. The seeds are surface-sterilised with a dilute solution of bleach and then sown on to the surface of the nutrient medium. Under controlled conditions of temperature and light, the seeds should germinate within a few weeks.

**Figure 9.7**

Asymbiotic germination of orchid seed uses agarose gel containing sugar, minerals, vitamins and organic supplements, such as pineapple juice, which contain a range of ingredients necessary for plant growth. This medium will supply a comparable range of nutrients to those provided by fungi.

a

b

**Figure 9.8**

(a) Large numbers of genetically identical orchid plants (clones) can be produced by vegetative micropropagation in the horticultural industry, thus making these highly sought-after plants available to a wide market and protecting the increasingly threatened wild populations.

(b) The orchid *Anacamptis morio* has been reintroduced to the wild after micropropagation.

## Orchid success stories at the Conservation Biotechnology Unit, Kew

**Figure 9.9**
The lady's slipper orchid (*Cypripedium calceolus*).

The lady's slipper orchid (*Cypripedium calceolus*) is Britain's rarest orchid. In the nineteenth century the British population was over-collected and only a single living specimen survived. From this, several thousand seedlings have been grown in the Conservation Biotechnology Unit at Kew and the species has been successfully re-introduced into the wild.

Kew has also worked with the University of Basel to propagate ten species of rare orchid. These have been taken back to Switzerland for re-introduction along road verges, which tend to provide undisturbed habitats.

Madagascar is one of the world's biodiversity 'hotspots' and has a great number of endemic orchid species. Many species are endangered due to human activity such as logging. Nine species of orchids have been returned to Madagascar as part of a conservation programme. A cryopreservation technique, similar to that described earlier for embryos, has been developed for a critically endangered Madagascan orchid called *Paralophia epiphytica*. It involves freezing orchid plant tissue in liquid nitrogen, which allows the tissue to be stored but subsequently defrosted and grown. This will enable living material of this species to be stored indefinitely, creating the potential for future reintroduction.

## 9.3 Micropropagation of palms

About 2,400 palm species (family Palmae/Arecaceae) are found in the tropics and subtropics, and many species of palm are under threat from over-collection or from habitat loss. Palms are one of the most economically important plant families, after the grasses and legumes, and provide many of the basic necessities of human life from food and timber to medicine and writing materials. Commercially important palm products include palm oil, coir fibre, carnauba wax, rattans and true sago. Because palms are so useful, many wild populations are overexploited and some species have become almost extinct. It is important to preserve the biological diversity of the different types of palm, as their potential uses may only become clear in future years.

In islands like Madagascar and Mauritius some endemic Arecaceae are critically endangered and facing extinction. If the biodiversity of this family is to be maintained then seed conservation is essential. Micropropagation techniques are also needed to produce specimens that can be grown on and planted out.

**Figures 9.10**

(a) The date palm *Phoenix dactylifera* was cultivated over 8,000 years ago in Babylon.

(b) The date fruit is the staple food of many people living in the semi-arid areas of the Arab world and wild relatives of the date are still found throughout this area. Besides being used as a source of food, the date palm also provides fibre, fuel and timber.

a

b

## Micropropagation of the bottle palm: a success story

**Figure 9.11**
The bottle palm, *Hyophorbe lagenicaulis*.

The bottle palm, *Hyophorbe lagenicaulis*, is restricted to Round Island, a small island 22 km north of Mauritius in the Indian Ocean. The introduction of goats and rabbits to the island during the nineteenth century caused a dramatic decline in the number of bottle palm plants. In 1996 the International Union for the Conservation of Nature (IUCN) organised a survey of the island and reported fewer than ten wild plants. In the wild, seed germination of bottle palms takes 5–6 months and is sporadic.

Scientists at Kew conducted studies to germinate seeds using micropropagation and used the seedling materials to produce somatic embryos – embryos developed from vegetative tissues. Unlike embryos from bottle palm seeds, these somatic embryos can be dried and then stored in liquid nitrogen. Creating somatic embryos involves cutting three-week-old seedlings into two longitudinal sections. The seedlings then develop somatic embryos on their cut surfaces. These embryos can either be grown on to form plants using micropropagation or else can be stored using liquid nitrogen. A protocol was developed to induce these groups of cells to develop into plantlets that can be transferred to soil. These techniques are being used to save *Hyophorbe amaricaulis*, a very close relative of the bottle palm, which was surviving as a single specimen in the Curepipe Botanic Garden in Mauritius.

## 9.4 Micropropagation of high-value plants in agriculture

Individual oil palm plants with desirable characteristics such as high yields, disease resistance or drought tolerance are cloned using micropropagation to produce many identical plants. When grown in a plantation, the palms have a uniform size and appearance and a greater chance of a guaranteed crop yield. Like other monocultures (areas containing just one single plant type), plantations of cloned palms have a particularly detrimental effect on local biodiversity and on wild populations of plants and animals. However, the high yields from the cloned monocultures mean that less land is needed to produce oil, leaving more available for other crops.

### Problems with bananas

Edible bananas and plantains are, in financial terms, the fourth most valuable food in the world, after rice, wheat and milk. The cultivars we eat today are seedless, and therefore sterile. It is thought that many thousands of years ago, two seeded bananas *Musa balbisiana* and *Musa acuminata* cross-bred and produced sterile hybrids. Seeded bananas are so densely packed with seeds that they are not edible.

Bananas are monocots and naturally produce offshoots from the base of their stems, forming genetically identical clumps. The seedless banana has survived because humans have tended and cultivated the plants and exploited their ability to reproduce asexually or 'vegetatively'. However, a lack of sexual reproduction results in minimal variation within the genetic material, hence banana clones are at particular risk from attack by many fungal, bacterial and viral diseases.

**Figure 9.12**

The African oil palm, *Elaeis guineensis*, is the highest yielding of all tropical oil-producing plants. Palm oil is pressed from the orange flesh of its fruits and palm kernel oil is pressed from its seed. Most palm oil is used in margarine, ice-cream, coffee whitener and other food products, but some is also used as animal feed and in the manufacture of soaps, lubricants and candles. However, the extensive use of palm oil is threatening to cause an ecological disaster as rainforest habitats are removed to allow oil palms to be grown.

**Figure 9.13**

Banana plant (*Musa* cultivar) showing the basal side shoots from which they are propagated.

One banana cultivar, 'Gros Michel', was wiped out by Panama fungal disease. The Cavendish cultivar is now the most traded banana. The future of the Cavendish banana is seriously threatened by black sigatoka leaf spot – a fungal disease that attacks the leaves and causes a greatly reduced yield of fruit. Control is by frequent applications of fungicide, increasing the growing space for each plant and improving the drainage of the plantation. All these measures are expensive and often beyond the means of subsistence farmers growing their staple crop for survival not profit. Micropropagation is therefore a viable option under these circumstances, especially since bananas have a short life. The technique involves cutting out the growing tip of the shoot, called the meristem, which is tissue that has not had a chance to be infected by the fungus. The cells of the meristem can be micropropagated and can form lots of new individual plants that are fungus-free. In fact, millions of banana plants are produced in this way every year; in India alone, over two million plants are propagated annually.

Micropropagation is one of the most useful and popular techniques available to plant scientists worldwide. With an ever-increasing human population, micropropagation is at the centre of research both for the conservation of biodiversity and to provide new cultivars to feed the world via its role in the genetic modification of plants.

# 10 Genetically modified plants

In the second half of the twentieth century, plant breeding and changing patterns of cultivation alleviated hunger in much of our world. However, as discussed in Chapter 2, the challenge of feeding an ever-increasing population in a sustainable manner on a planet of finite resources remains. Indeed, this challenge has recently been thrown into sharper focus by the potential for crop failures in a changing climate, and by the increasing use of agricultural land for growing biofuels, covered in Chapter 4. Despite the challenges facing the agricultural industry, crop yields have continued to rise. The question now is, can these rises be sustained?

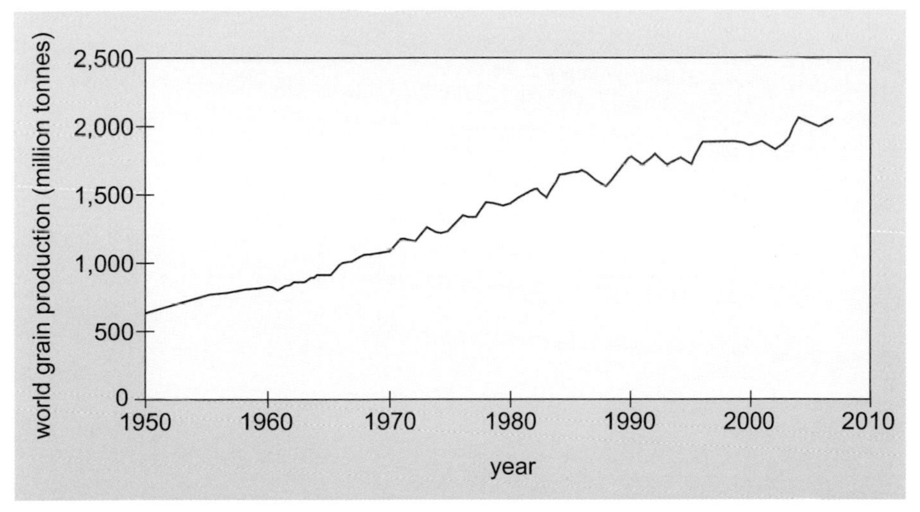


**Figure 10.1**
World grain production has increased steadily since the 1950s, and the producers have largely been able to keep pace with demand. This is in part due to the 'Green Revolution' in which new high-yielding varieties have been bred specifically to respond particularly well to high inputs of fertilisers.

Scientists believe that in some parts of the world farmers are close to producing the maximum yield that can be obtained from existing crop cultivars, so it is important to look at other methods that could allow continued increases in crop production. One strategy has been the development of genetically modified (GM) plants, as these may give increased yields, but the application of this technology has opened a wide-ranging and global debate. On the one hand, supporters of GM plants argue that the technology is a vital tool in plant breeding which can address many of the future challenges facing agriculture and could allow increased yields and more nutritious crops to be produced, particularly for those living in developing countries. On the other hand, opponents argue equally strongly that GM plants pose unacceptable environmental, ecological and economic risks, and that it is unethical to grow them. Some of these risks include GM crop plants breeding with wild relatives and the 'unknown' effects of introducing GM plants into the environment. At the time of writing (2009), some countries, Americas and parts of Asia, for example, consider this is a risk worth taking. For others, including European countries, it is not. This chapter describes the science of the technology and illustrates some of its current and proposed uses. It does not enter the debate on GM plants, nor does it attempt to influence its outcome.

## 10.1 History of plant breeding

Plant breeding is the manipulation of plant genes for human requirements; this may be to create crops that produce higher yields or have disease resistance, or to produce ornamental plants for the horticulture industry. For thousands of years, traditional methods have resulted in myriad forms of crops adapted to different uses and to local conditions.

**Figure 10.2**

Soybean (*Glycine max*) is the most widely grown GM plant in the world today.

### What is a gene?

As discussed earlier, DNA comprises much of our own and plants' genetic material. In more detail, DNA (deoxyribonucleic acid) is a polymer consisting of two strands, which are held together by weak chemical bonds. The strands of DNA are carefully packaged to fit into the nucleus of a cell and at certain points in the cell's life cycle are visible under the microscope as sausage-shaped structures known as chromosomes. The number of chromosomes is specific for most species: humans, for example, have 23 pairs of chromosomes and beans (*Phaseolus* species) have 22 pairs of chromosomes.

Genes are the segments of DNA that produce the proteins which control particular characteristics or traits. For example, the white colour of a pea flower petal is due to the presence of a 'white gene' or the absence of a gene for colour. The observable physical characteristics make up the 'phenotype' of the organism. A gene has a sequence of chemicals (bases) within its DNA and the genome of a species is the entire DNA sequence within that species and includes all the genes. For instance, the rice genome is thought to contain around 45,000 different genes.

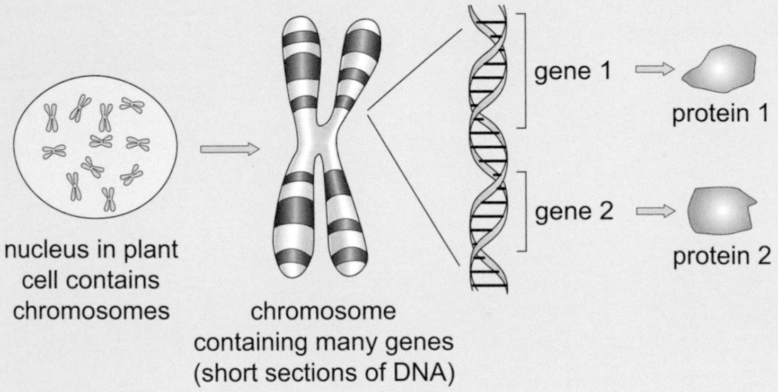

Cells of different species may contain different amounts of DNA but every cell of a particular plant will contain exactly the same sequences of DNA. This is quite remarkable when the cell types differ so much within a plant: some form root cells, some form leaf cells and others form petals. This 'specialisation' of cell types is due to the control mechanisms within the cell, which 'switch' certain genes on and off. So, a gene for a white pigment would usually only be switched on in the petal cells, for example. When a gene is switched on, the sequence of bases in the DNA determines, or 'codes for', the protein to be produced. Spontaneous changes in the DNA sequence can occur and these are known as mutations; they can have serious effects on the activity of the gene or be of no consequence to the activity. The gene is referred to as being expressed, or being 'active' or producing an 'active product', when the protein is actually synthesised from the gene.

A 'transgenic' or 'GM plant' is one in which a foreign gene, possibly from another species, has been artificially introduced into the genome. This may result in an altered phenotype of the plant, in which case its presence would be clearly observable, or it may not alter the appearance of the plant at all; in either case the genome of the plant is changed.

nucleus in plant cell contains chromosomes
chromosome containing many genes (short sections of DNA)
gene 1
protein 1
gene 2
protein 2

**Figure 10.3**

A schematic diagram showing DNA packaged within chromosomes in the nucleus of a plant cell, the location of genes on the chromosomes and how the information encoded within the genes is used to make different proteins.

**Figure 10.4**

Many variegated plants have been produced by cross-breeding plants that have natural mutations. The mutated characteristic can often revert back, as demonstrated by variegated holly reverting to white.

Plant breeders have managed to develop cultivars with plant traits that have arisen through mutations. These mutations can produce interesting plants but they arise infrequently and are often unstable, so the plant may revert back to the original type. For example, variegated leaved plants can produce a fully white or fully green offshoot or leaf. By using asexual breeding techniques, such as micropropagation and the taking of cuttings as described in Chapter 9, some of these mutations can be maintained.

Advances in plant breeding arose through the discovery and understanding of genetic inheritance first recognised by a monk, Gregor Mendel, who devised a series of breeding experiments to investigate different characteristics in peas. These experiments led to the principles of Mendelian genetics and from these the systematic techniques of creating hybrids were further developed.

**Figure 10.5**

(a) Gregor Mendel (1822–1884) was a monk in what is now the Czech Republic. He was instrumental in working out how characteristics could be passed on from one generation of plants to the next.

(b) Mendel's copy of his *Experiments in Pea Hybridisation* (1866) (left), describes his discovery of the statistical laws of heredity.

a

b

## 10.2 Selective breeding

After Mendel's work, the inheritance of desired characteristics had a greater scientific basis and selection could be directed with greater precision. Desirable characteristics can be cross-bred between different varieties, either cultivated or wild ancestors, through transfer of pollen from one parent flower to the stigma of a flower from the second parent (see Figure 5.9). The resulting offspring of the cross is referred to as a hybrid and contains a mixture of useful and less useful genes. To increase the purity of the genetic make-up of the resulting plants, the daughter plants can be 'backcrossed', which involves crossing offspring with genetic material of one of the parents until the desired characteristic is stable and maintained through generations.

a

b

c

**Figure 10.6**

(a) The Monterey cypress (*Cupressus macrocarpa*) is renowned for its ability to survive harsh environments such as the sea air.

(b) The Nootka cypress (*Xanthocyparis nootkatensis*) is a fast-growing species.

(c) Hybridisation between the two resulted in the Leyland cypress (x *Cuprocyparis leylandii*), which has inherited both of these characteristics and can sometimes be considered to be a nuisance if planted in inappropriate locations.

Hybridisation usually occurs between plant species which belong to the same genus and the resultant progeny is known as an 'interspecific hybrid', for example the hybrid larch (*Larix* x *marschlinsii*). This hybrid is a popular tree in the forestry industry as it puts on early growth at a greater rate than either of its parents, the European larch (*Larix decidua*) and Japanese larch (*Larix kaempferi*). More rarely, a hybrid can be created between members of the same family but from different genera. These are referred to as 'intergeneric' hybrids and the Leyland cypress (x *Cuprocyparis leylandii*) is one example, which, like many hybrids, is sterile.

### Limitations of traditional plant breeding

Traditional plant breeding and cultivation techniques have been successful methods for developing new varieties and cultivars but they are not without limitations. First, cross-breeding cannot generally occur between different taxonomic families, so the number of possible crosses is restricted. If a particular plant, for example, a broccoli cultivar, demonstrates good disease resistance, that characteristic cannot then be bred into a plant such as rice. This is because pollen has natural barriers and signals that ensure that it can only pollinate a plant of its own or a closely related species. This means that no pollen, and hence genetic material, is lost to an inappropriate host plant. Second, during breeding programmes a whole mass of genes is expressed in the progeny plants. It is therefore difficult to obtain and isolate just one desired character without changing other aspects of the plant. Finally, breeding programmes take years to reach the point when a desired trait is stably maintained within a plant population. For example, it can take 25 years to breed a new reliable variety of pear. It also requires a lot of space and resources, and hence breeding programmes are often restricted to research institutes with large areas of land.

## 10.3 What are GM plants and why might they be useful?

GM plants are those that contain genetic material (DNA) that has been transferred to them by means other than natural traditional breeding methods. This is perhaps central to understanding why GM plants are radically different. Traditional breeding has tended to produce new cultivars that have lost a gene function, whereas GM plants tend to have extra genes added and these added genes may come from other species. In GM plants this 'additional' DNA typically resides in the nucleus of the cell; however, it can be present simply in the cytosol (the area of the cell outside of the nucleus) as a self-multiplying entity such as a virus (an infectious agent, smaller than a bacterium, with its own genetic material).

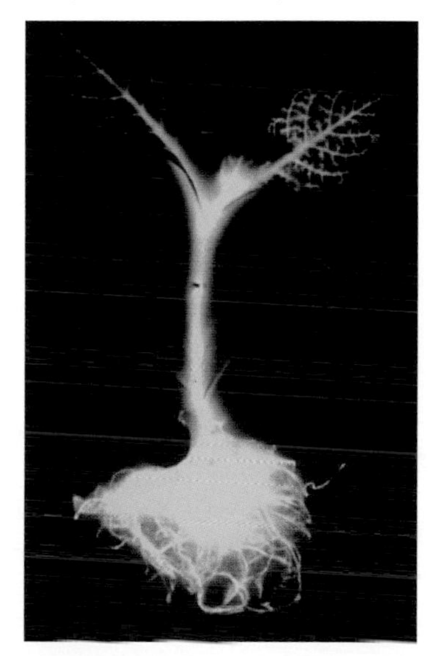

**Figure 10.7**
The clearly observable result of the technology behind producing GM plants: a tobacco plant that has been genetically modified to contain the gene that allows a firefly's tail to glow in the dark. It clearly shows where the foreign gene is active. However, most GM plants are not obviously different from non-GM plants.

The notion that GM plants could enable the introduction of new characteristics by a quicker and more specific process than traditional conventional breeding is an attractive one to some people, although not to everyone. There is also the real possibility of using genetic modification to introduce non-plant genes, which would allow novel, useful characteristics to be expressed within plants. For example, it should be possible to produce: avocados that are soft and ready to eat when you buy them and do not subsequently rot; crops with increased yields, or resistance to pests and diseases; pears or tomatoes that do not drip juice when you eat them; or trees whose wood can be easily pulped to reduce the economic and environmental cost of paper-making. Many ideas have been suggested over the years, often sparking headlines in the media. Early proponents of GM plants suggested modifying non-legume plants so that they could fix their own atmospheric nitrogen, thus reducing the need for fertilisers and thus the potential for nitrogen pollution. Later ideas included plants which could withstand frost because they had been modified to contain a fish 'antifreeze' protein, or grass that could glow in the dark offering golfers the opportunity to play their game long into the night. Whether or not the idea is reasonable, or apparently outlandish, the application of the technology behind GM plants is only of worth if it adds to the real value of the plant or crop in question.

## 10.4 Techniques for producing GM plants

Techniques have been developed to produce two general types of GM plant, each with its own particular benefits:

- those in which the plant's own genome has been modified
- those into which a virus has been introduced that has been genetically modified with the addition of the gene of interest before insertion into the plant.

### Genetic modification of the plant genome

In GM plants where the plant's own genome has been modified, the new 'foreign' gene (or set of genes) will become stably integrated so that it will be passed on to offspring (inherited). This means that the characteristic can be introduced via a GM plant into traditional breeding lines and practices. Modifying a plant's genome can be carried out either by using *Agrobacterium tumefaciens*, a bacterium found within the soil modified to transfer foreign DNA into the plant, or by a process known as particle bombardment.

#### An improbable relationship between a bacterium and a plant

At the turn of the twentieth century, two workers at the US Department of Agriculture were working on tumours, or galls, that formed on Paris daisy (*Argyranthemum frutescens*). The tumours were similar to those found on peach trees in orchards. The tumours formed at the base of the plant and appeared to originate from the plant itself. Over two years, Smith and Townsend showed that bacteria isolated from the tumours could induce further tumours when inoculated by pinprick onto an otherwise healthy plant. When they published their results in 1907, they called the bacterium responsible *Bacterium tumefaciens*. This soil bacterium is found widely in temperate climates and is now known as *Agrobacterium tumefaciens*. You may be familiar with the tumours it forms as they are known as crown gall and are often seen on trees.

**Figure 10.8**
Crown gall tumours formed on walnut trees by *Agrobacterium tumefaciens*.

Smith, in particular, was fascinated because the tumours that formed on the plants resembled tumours seen in animals; the infected cells apparently multiplied without control. Moreover, it was subsequently demonstrated that the tumours, once initiated, could continue to grow in the absence of the *Agrobacterium*. This suggested that the change that had occurred was stable in the cells.

In the 1930s it was found that a dried extract from the medium in which *Agrobacterium tumefaciens* was grown could induce explants to form tumours. The active substance responsible was indole-acetic acid (IAA), which was later identified as the plant growth regulator auxin. It seemed that the bacterium triggered cell proliferation by the production of a plant growth regulator, yet this growth could be maintained in the absence of the bacterium.

Like many bacteria, in addition to its single chromosome, *Agrobacterium tumefaciens* contains a circle of double-stranded DNA called a plasmid. While maintaining a plasmid may provide the host bacterium with some advantages, this comes at a cost to the cell in terms of the energy required to produce and maintain the plasmid. Hence, the host bacterium may lose the plasmid when it grows in stressful conditions. Being a soil bacterium, *Agrobacterium tumefaciens* prefers temperatures of 20–30 °C; if grown at 36 °C it loses both the plasmid and the ability to trigger tumour growth. The two are apparently linked and the plasmid became known as the tumour-inducing, or Ti, plasmid. This raised an interesting question: if the presence of the Ti plasmid in *Agrobacterium tumefaciens* is required to induce tumours, why do tumours continue to grow in the absence of the bacterium? Subsequent work revealed a surprising answer.

During infection of a plant by *Agrobacterium tumefaciens* only a specific portion of the Ti plasmid, the transferred DNA (T-DNA), is transferred from the bacterium to the infected plant cell. The T-DNA travels to the nucleus of the host plant cell and integrates into (becomes part of) the genome, where it is stably maintained. Once within the plant genome, genes carried on the T-DNA become active and produce enzymes that in turn produce the auxin plant growth regulators in the infected cell. These act to short-circuit the host plant's control of cell growth and proliferation, and thus trigger tumour growth.

**Figure 10.9**

*Agrobacterium tumefaciens* can be used to produce GM plants as it contains a tumour-inducing (Ti) plasmid, which contains both virulence (*vir*) genes and a transfer-DNA (T-DNA) region. The bacterium attaches to a plant cell, and the T-DNA and Vir proteins are transferred to the plant through a transport channel. Inside the plant cell, the Vir proteins promote the integration of the T-DNA into the plant genome.

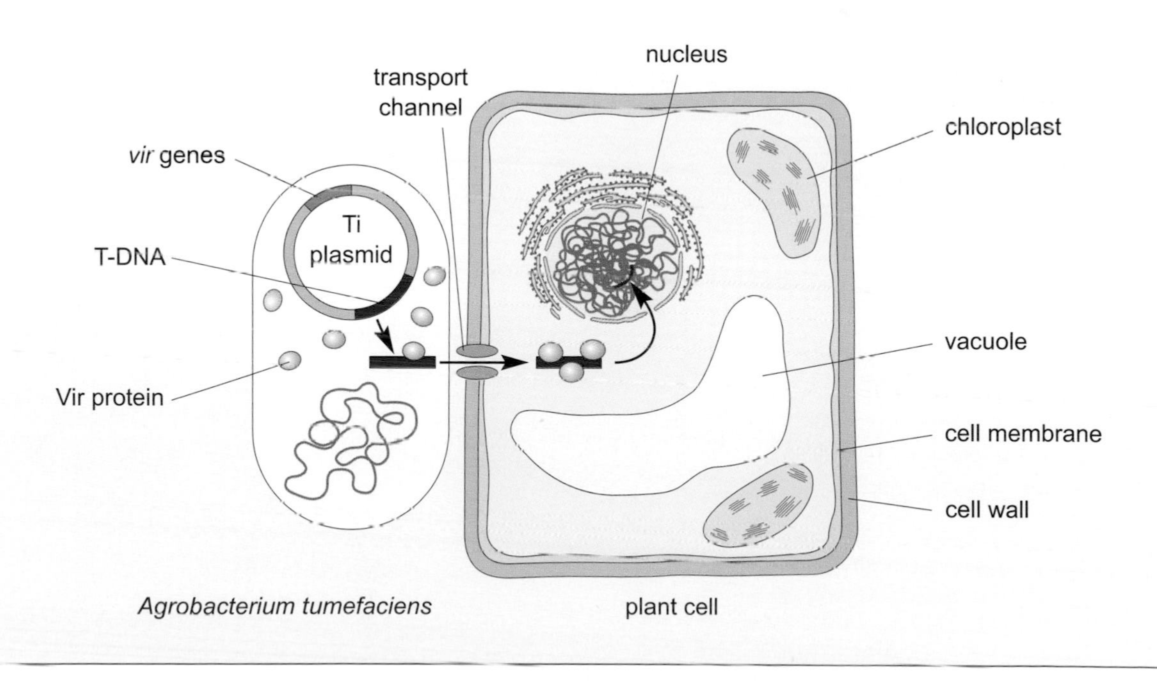

Research into the natural gene transfer system of *Agrobacterium tumefaciens* determined that it could transfer DNA obtained from any species. Moreover, the genes responsible for tumour growth could be removed without affecting the T-DNA transfer process. Hence methods have been devised to allow foreign DNA to be transferred into *Agrobacterium tumefaciens* within the T-DNA so that it can be subsequently inserted into the plant genome.

Particle bombardment overcomes the need to use *Agrobacterium tumefaciens* in the gene transfer process. Here, the DNA to be transferred into the plant genome is coated onto inert microscopic particles of gold or tungsten. These are shot into the target plant tissue, using a particle gun, in a process analogous to using a shotgun. The particles carry the DNA into the cell and ultimately the DNA dissolves from the particle and is integrated into the plant genome. Though the particle gun is effective in transferring DNA to the plant cell, care is needed not to blow the target tissue to pieces!

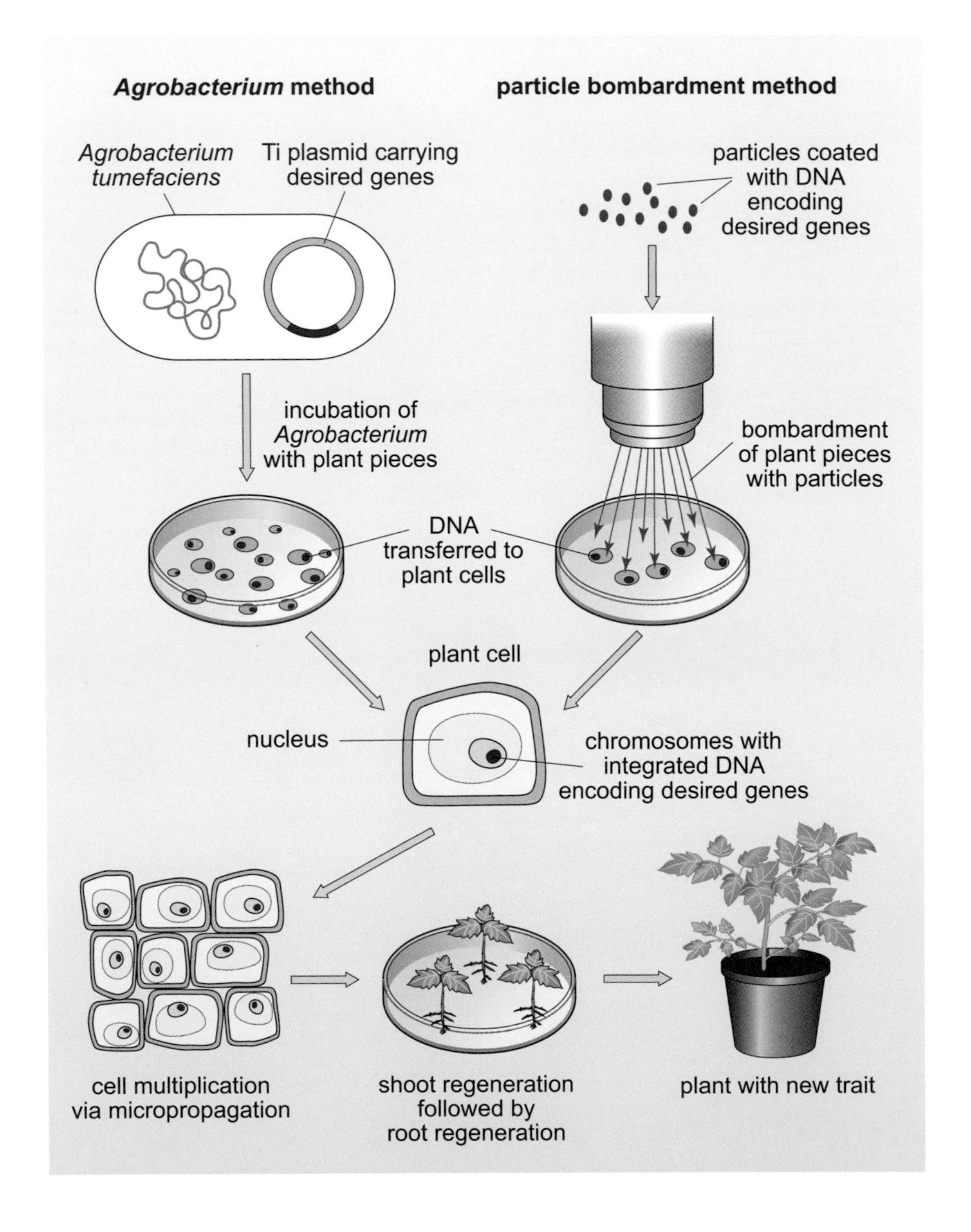


**Figure 10.10**
Creating a GM plant using *Agrobacterium tumefaciens* or particle bombardment.

Gene transfer by both *Agrobacterium tumefaciens* and particle bombardment is a relatively inefficient process and scientists need to be able to identify the modified cells that contain the transferred foreign DNA. To enable them to do this, the DNA transferred to the plant cell usually contains a gene providing resistance to an antibiotic. If cells contain this gene, they are able to grow and proliferate in the presence of the antibiotic. If not, the cells are unable to proliferate. Hence, following treatment with *Agrobacterium tumefaciens* or particle bombardment, the newly growing cells are micropropagated in the presence of the antibiotic so that only cells containing the foreign DNA can grow.

### Genetic modification by viruses

GM plants created by the introduction, or inoculation, of viruses cannot be employed in traditional breeding methods because the new gene is not inserted into the plant genome but is maintained outside the nucleus in the cytosol of the cell and is likely to be lost during reproduction. However, GM plants modified by viruses have the potential to make large quantities of the protein whose encoding gene has been inserted into the viral genome. This raises the possibility of growing plants infected with viruses to produce personalised pharmaceutical proteins to treat individuals; for example, antibodies specifically designed to counter the tumours of a lymphoma patient. Moreover, the DNA of the virus can be modified to include other foreign genes before the virus is inoculated into the plant cell in a relatively quick and straightforward manner, so that insertion of foreign DNA into plants may take only a matter of days. There are restrictions on the size of the gene that can be inserted into the viral genome. However, work on this technique is continuing and it is likely to become of greater significance in the future.

## 10.5 Examples of GM plants

It is important that plant scientists can control where in the plant a particular transferred gene is both present and active. For example, if the gene provides resistance to an insect pest, it might make sense for it to be active in the leafy parts of the plant. If the gene provides resistance to nematodes (pests found in the soil) activity in the root might be preferred. On the other hand, if the interest lies in modifying starch synthesis in potatoes, the gene needs to be active in developing tubers.

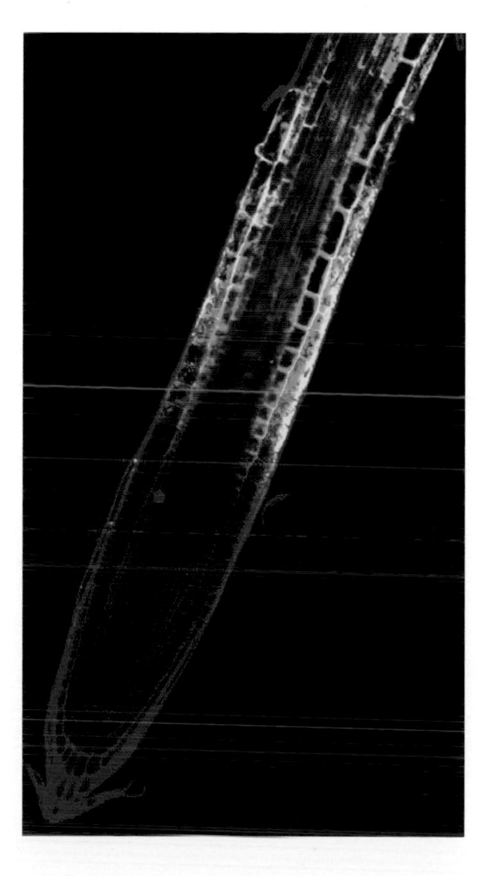

**Figure 10.11**
The activity of a transferred foreign gene can be analysed by obtaining plant material and then visualising where the gene is present and active. In this example, the scientists were studying the genes that control the elongation of cells, a process that occurs just behind the growing tip of the root and the shoot. The red colour indicates that the inserted foreign gene is being expressed.

## Herbicide-resistant GM crops

Soybeans account for over half the area of GM crops grown worldwide. Monsanto's Roundup Ready® soybean is the best-known cultivar. GM technology has been used to make the crop resistant to the non-selective herbicide Roundup® (also marketed by Monsanto), which contains the active ingredient glyphosate. The gene that gives resistance was transferred from a strain of *Agrobacterium*. The benefit for the farmer is that weed control is easier, as there is less likelihood of the herbicide damaging the crop plant. Studies have shown that the volume of herbicide used by farmers stays the same or decreases slightly, but fewer different types are used, thus lowering costs and potentially reducing the impact on the environment. Use of Roundup® also makes low-tillage (using a minimum of farm machinery) cultivation easier, sometimes allowing the cultivation of a second crop of soybean in the same growing season. The Roundup Ready® package of crop and herbicide is now available for other crops, including maize, sugar beet and oilseed rape.

## GM plants resistant to insects

An early promise of GM plant technology was to take genes from non-plant sources and use them to create novel, beneficial characteristics in plants. An example of where this promise has been realised is in the use of the gene for a toxin from the bacteria *Bacillus thuringiensis*. The toxin produced is called the Bt toxin.

Bt toxin is a protein that crystallises (solidifies) when the bacterium forms spores. When insects ingest the spores, the toxin is digested to produce the active toxin, which kills the insect. The toxin only has its effect when digested in the gut, and different strains of *Bacillus thuringiensis* synthesise proteins that act against different specific insects. Hence, the Bt variety *kustaki* is active against Lepidoptera (butterflies and moths) and the variety *israelensis* acts against Diptera (mosquitoes and blackfly). Preparations of *Bacillus thuringiensis* spores have been marketed for some time as a pesticide spray but the activity of this spray proved unstable in field conditions. Hence, a desire for the gene for the toxin to be inserted into susceptible plants made it an early target for GM plant technology.

a b

**Figure 10.12**

Comparing the effects of insect infestation in Bt corn (maize) (a) and non-Bt corn (b). The corn cobs without the Bt gene show severe feeding damage caused by the Mediterranean corn borer, *Sesamia nonagrioides*. Mould has entered the cob at insect feeding sites and further reduced cob quality.

GM maize (corn) and cotton crops that have been modified to contain the Bt toxin have been particularly successful. These GM crops have been widely adopted by farmers in the USA, where it has been estimated they have significantly reduced the cost of production because of lower insecticide applications and increased yields. Bt cotton has also been widely grown in both India and China where the majority of the growers operate on small farms and are relatively resource-poor. It is estimated that in China in 2008 there were 7.1 million growers who fall into this category, and in India, 5.0 million.

A danger of deploying GM Bt crops is that over time the insects develop resistance to the toxin. To prevent this, in the USA, farmers are required to plant 20% of the crop as traditional non-Bt plants, providing a refuge for the insects. This 20% must be adjacent to the Bt crop, say in rows or as a border. The refuges provide habitats for the non-resistant insects, thereby reducing the selection pressure for resistant insect selection.

**Figure 10.13**

Harvest-ready crop of GM cotton (*Gossypium hirsutum*). Photographed In Mississippi, USA, where GM cotton was introduced in 1996.

## GM plants resistant to viruses

Virus cross-protection is the idea of protecting a plant from viral infection by prior infection with a related virus that produces milder symptoms. It arose in the late 1920s and was based on the observation that inoculation of tobacco plants with a tobacco mosaic virus (gTMV) that produced green mosaic symptoms prevented another strain of TMV (yTMV) from producing its normal yellow mosaic symptoms. Thus, cross-protection can be thought of as the ability of one virus to suppress, or delay the appearance of, symptoms caused by a later 'superinfecting' virus.

Virus cross-protection has been used in agriculture but its success can be limited, not least because of the practical difficulties of inoculating a crop with a virus. However, the phenomenon sparked early interest in producing a GM plant containing gTMV genes and testing whether the GM plant then resisted yTMV infection. The resultant GM plants were indeed resistant to inoculated yTMV and this opened up the prospect of using this methodology for a range of other viruses that have a similar genetic make-up to each other but differ in the severity of the symptoms they generate.

## Papaya Ringspot Virus (PRSV)

**Figure 10.14**
PRSV-infected papaya field.

A notable example of the success of virus cross-protection has been the control of papaya ringspot virus (PRSV). Papaya is a major fruit crop in South and Central America, Asia and Hawaii. There are two types of PRSV: PRSV-p infects papaya and cucurbits, such as melon and squash, and PRSV-w infects only the cucurbits. In papaya, PRSV interferes with the plant's ability to photosynthesise. This results in stunted, chlorotic plants (plants which are pale due to a lack of green chlorophyll) with deformed inedible fruit, and eventually leads to the plants' death.

PRSV was first identified on the Hawaiian island of Oahu in the 1940s. In the 1960s it began seriously to threaten commercial papaya growing, so the industry moved to the virus-free island of Hawaii. However, by the late 1990s the virus had reached Hawaii, and with production dropping by almost 40%, the industry there faced collapse.

In the late 1980s scientists at Cornell University who had been working on virus cross-protection adopted a similar tactic to that used in tobacco to protect against TMV. They used the particle bombardment approach to transform papaya with a gene for one of the proteins found in the outside coat of PRSV. One line, derived from the true-breeding commercial cultivar 'Sunset', proved in glasshouse trials to be resistant to infection with PRSV HA, a severe strain of the virus from Hawaii, and this was selected for field trials on Oahu.

**Figure 10.15**
GM papaya with the PRSV HA gene demonstrating resistance to PRSV HA (left). Papaya without PRSV HA (right).

**Figure 10.16**
A GM papaya field trial on Hawaii. The green GM variety 'Rainbow' is on the right and the non-GM PRSV-infected chlorotic plants are bottom left.

## GM crop plants

Rice (*Oryza sativa*) is a staple food for millions of people, particularly in the developing world. In 2000, rice, which does not normally produce β-carotene, was genetically modified to produce the orange pigment. This new rice was given the name 'Golden Rice'.

In 2005, a new rice, 'Golden Rice 2' was developed which produces 23 times more β-carotene than the original Golden Rice. Golden Rice was designed for human consump-tion to address vitamin A deficiency (β-carotene is a precursor of vitamin A). Up to 2 million people a year die and over 200,000 people become irreversibly blind as a result of vitamin A deficiency. There has been considerable difficulty in using Golden Rice as a source of dietary β-carotene, due in part to scientific concerns over the safety and efficacy of such a genetically modified food source. Additionally, the companies funding the initial research hold patents that cover the technology used to produce GM rice and can therefore charge significant sums for the seed. However, Humanitarian Use Licenses have been granted for the growth of Golden Rice by farmers as long as they do not make more than US $10,000 a year from sales of the rice, although it should be noted that Golden Rice is not yet under cultivation. Nutritionists have raised concerns over whether the β-carotene content of Golden Rice would translate into increased consumption of vitamin A for those who eat it.

a

b

**Figure 10.17**
Normal rice (a) and Golden Rice (b).

### GM plants and the pharmaceutical industry

As well as GM crop plants, GM plants or their cells can be used to make any proteins, including those of interest to the pharmaceutical industry, such as antibodies and vaccines. GM plants may provide a cheaper means of producing high value proteins but, to be accepted, they need to be grown under conditions where no cross-contamination of other plant products takes place, and the pharmacological protein in question needs to be isolated in a pure form with ease. One way of partially overcoming these problems is to produce the proteins in cultured plant cells. At time of writing (2009), a company in Israel is testing the idea of using carrot cells cultured in giant plastic bags to produce an enzyme (a type of protein) used in the therapy of Gaucher disease (a genetic disease in which fatty lipids accumulate in cells and certain organs).

## 10.6 The current position regarding the growing of GM crops

Most of the opposition to GM crops relates to concerns that there may be a higher risk in consuming GM food than in consuming non-GM food. In addition, the unknown risks posed by a GM plant breeding with related wild species and the modified gene 'escaping' into non-crop populations is also cited as a concern. The fact that the industry is dominated by a small number of private companies who up to now have generally chosen to introduce traits into crops for profit rather than need is also cited by some as a reason why the technology needs to be carefully monitored. Some people also believe that there is sufficient food to feed the world population, and that the problem is with its distribution. Organic farmers do not use GM crops, although paradoxically the use of GM can be an effective method of producing plants with pest resistance, which would result in reduced use of pesticides.

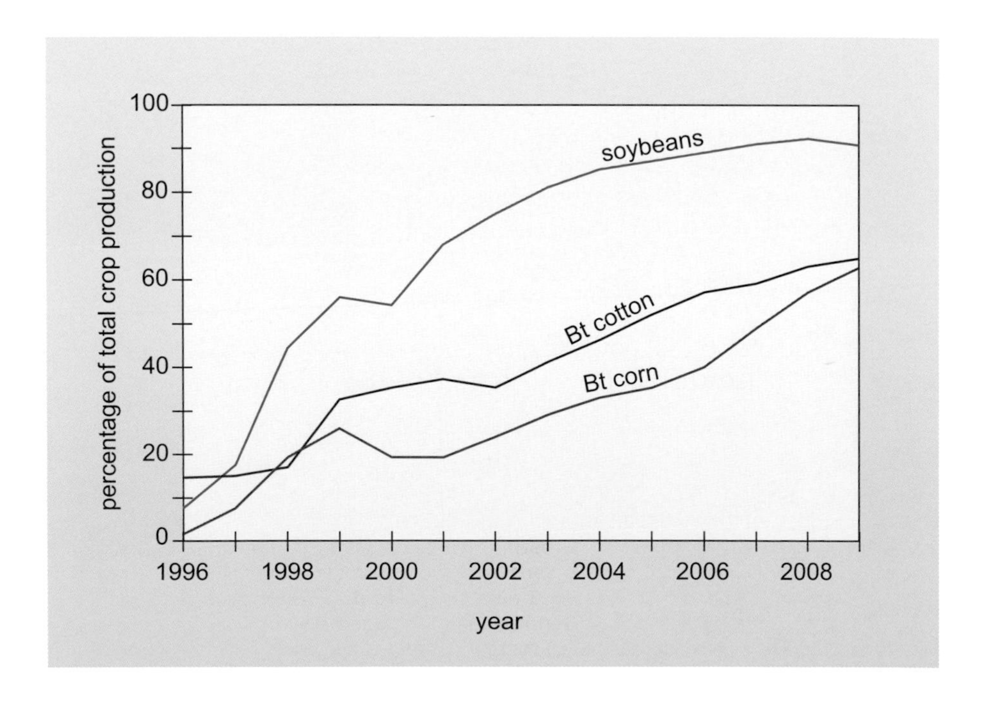


**Figure 10.18**
The United States of America has rapidly embraced the use of GM crops with well over 50% of the total of production of soybeans (*Glycine max*), cotton (*Gossypium hirsutum*) and maize (corn, *Zea mays*) being GM varieties.

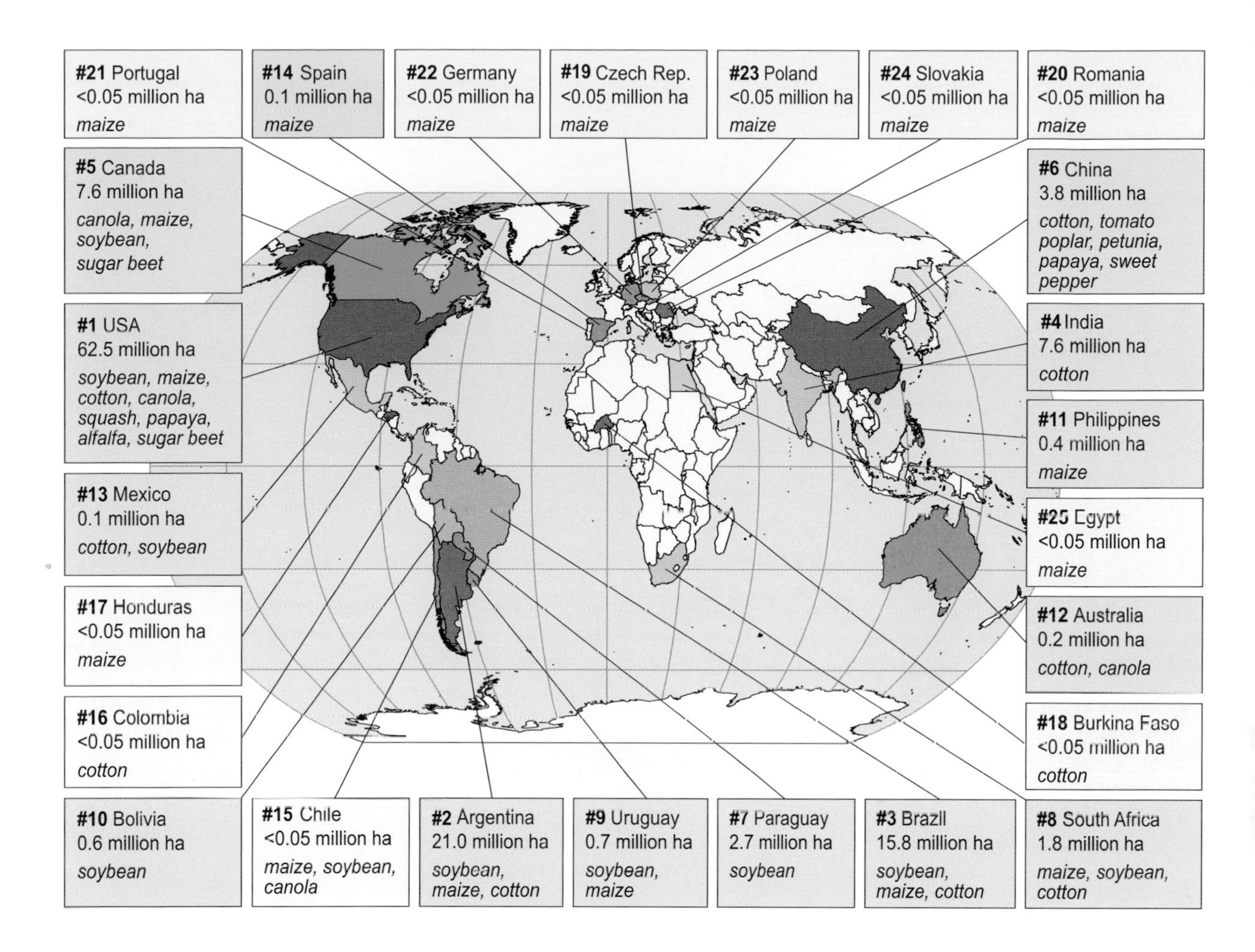


The total area of GM crops in cultivation has increased from 1.7 million hectares in 1996 to 125 million hectares in 2008. The main GM crops are soybean, maize, cotton and canola (oilseed rape). The best-known GM seed company is Monsanto; its GM seeds (some licensed to other companies) account for the majority of the world area devoted to genetically engineered seeds.

The use by farmers of plants that have been genetically modified has shown a steady increase in some parts of the world, although there are some areas, notably Europe and the UK, where consumer resistance has slowed or even prevented uptake of the technology. It will be interesting to see if, and how, the potential markets for plants that have been genetically modified change in the coming years.

**Figure 10.19**

Map showing the 25 countries where a total of 125 million hectares of GM crops were grown in 2008. This equated to some 8% of the total area of cropland used throughout the world. More than half (55%) of the global population lives in these 25 countries. Only 14 countries had more than 50,000 hectares of GM crops.

# 11 Natural plant protection

With the increasing human demand for plant crops, comes a greater need for maximum crop productivity. Protecting these crops from competing weeds, pests and diseases is essential to ensure this productivity is achieved. Concerns over the effects of some synthetically produced chemicals on the environment and on human health, along with evidence of increased resistance to pesticides and the banning of many active ingredients, has put pressure on the agricultural and horticultural sectors to reduce their use of some chemicals. The need to develop other methods of plant protection is greater than ever.

Throughout the history of agriculture, various techniques have been employed to protect crop plants and recent research has provided scientific evidence to support some of the traditional horticultural techniques.

Biological control is the control of pests by use of other organisms. There are a number of control systems that include deterring a pest from attacking the crop and encouraging natural enemies of the pest in question. Many of the biological controls require the natural balance of the ecosystem to be maintained, so there will still be some natural loss: the aim is to hold this loss at a manageable level.

a

b

**Figure 11.1**
(a) Growing onions, garlic or chives with carrots can deter (b) the carrot root fly (*Psila rosae*), which is a common pest to growers of root vegetables.

## 11.1 Biological control versus chemical control

Biological control has the benefit of being non-toxic and inexpensive because there is no need for the application, or indeed the reapplication, of pesticides or herbicides. There are many types of pesticides and herbicides and the ways in which they work vary. For instance, chemical pesticides tend to belong to three main groups. One of these groups, the organochlorine type, to which the now infamous DDT insecticide belongs, contains several members that function by affecting the nervous system of insects. Unfortunately, DDT is long-lasting, both in the environment and within organisms, and as it can be passed from one organism to another it can accumulate to high levels in animals at the top of food chains. The other two groups of pesticides are the organophosphates and the carbamates, both of which also work by affecting the nervous systems of insects. As organophosphates can affect the nervous system of mammals as well as insects, carbamates are the main pesticide of choice. Great care has to be taken to ensure that there is limited exposure to pesticides and herbicides for anyone who has to work with or apply such chemicals. These drawbacks to chemical control have led to a rise in the popularity of biological methods.

## 11.2 Use of naturally occurring predators

The principle behind biological control is to use predators that occur naturally to limit the levels of the plant pest. This can involve either direct predation or the use of parasites, which are insects that complete their life cycle, or part of it, within another organism, known as the host. The parasite can infect the adult pest, its eggs (destroying the pre-hatched nymphs), or its young (destroying the larvae).

### *Encarsia* wasps and whitefly

One of the first biological controls to be used commercially was the parasitic wasp *Encarsia formosa*. The wasp uses the whitefly larva to act as a host for its eggs, leading to the destruction of the whitefly larva when the adult wasp emerges following 28 days of development. This predator–prey interaction has been exploited by tomato growers who release commercially available *Encarsia* into the growing area. The *Encarsia* are supplied as parasitised whitefly larvae fixed on cards which are hung in the glasshouse ready for the wasps to emerge. Obviously there has to be a sufficient number of whitefly within the growing area to make releasing *Encarsia* economically worthwhile. The *Encarsia* themselves have fairly specific conditions which they prefer, but a glasshouse environment fulfils most of these criteria. There is also the problem of confining the *Encarsia*, although normally enough remain within the glasshouse to keep the whitefly numbers under control. Those that escape tend not to survive in the cooler conditions outside.

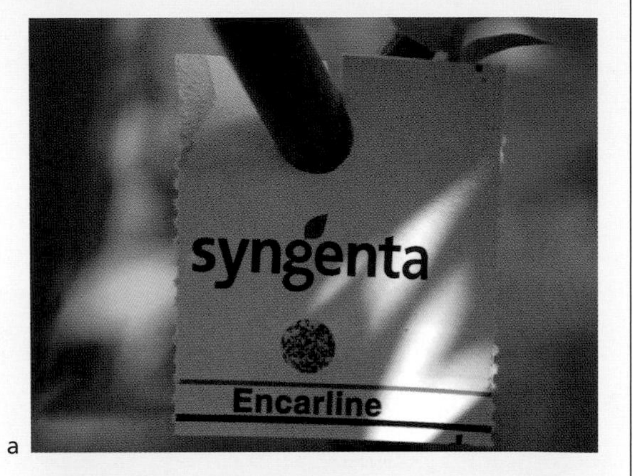


a

b

c

**Figure 11.2**

(a) Packets of commercially available *Encarsia* which infect the larval stage of the whitefly.

(b) An adult *Encarsia* wasp lays its eggs into a two-week-old whitefly larva.

(c) Once parasitised, the immature whitefly larvae turns black. The wasp grows and develops within the larva and, in 28 days, the adult wasp emerges out of the now-redundant cocoon.

Of the other insects that can be used in biological control systems, probably the most familiar is the ladybird (ladybug). This beetle is a consumer of a number of different aphids and it is possible to purchase these beetles commercially.

a

b

Figure 11.3

(a) Hoverflies and (b) butterflies are attracted into gardens by a variety of plants, especially goldenrod (*Solidago* species) and many of the species from the Umbelliferae (Apiaceae) and Compositae (Asteraceae) families found in traditional 'cottage garden' planting schemes, such as fennel (*Foeniculum vulgare*) and Michaelmas daisy (*Aster* species).

## Planting to encourage predators

Natural predators of pests can be encouraged into the garden or fields by creating natural refuges, habitats and 'set-aside' areas. Hoverflies, for example, are beneficial insects and prey on a variety of garden pests, particularly aphids. The adult hoverflies consume aphid honeydew, the sticky secretion emitted from adult aphids, and lay eggs nearby, while the hoverfly larvae are predators of aphids.

Providing shelters such as wood piles and straw can provide habitats for over-wintering lacewings and adult ladybirds (ladybugs) and their larvae, both of which are voracious predators of aphids.

Figure 11.4

Creating areas of wild plants, such as nettles (*Urtica dioica*) and ox-eye daisies (*Leucanthemum vulgare*), is useful for attracting natural pest predators. Early-flowering plants produce lots of pollen, which helps to support large populations of predatory insects and mites and enables their numbers to build up before the pest insects take hold.

a

b

**Figure 11.5**
To control aphid populations, it is beneficial to encourage adults (a) and larvae (b) of the 7-spotted ladybird (ladybug) into the garden.

The key to successfully attracting beneficial insects and wildlife is to have a high biodiversity of plant species. The greater the variety of plant species present, then the greater the diversity of insects and other predators that are attracted to the areas. From 1993, the UK and mainland Europe introduced the 'set-aside' of agricultural land, which allowed farmers to remove a small proportion of their land from cultivation in order to control agricultural over-productivity; they were compensated for their loss in income from the European Union. The policy of set-aside encouraged the improvement of wildlife habitats and therefore biodiversity, including the numbers of beneficial insects and over-wintering birds.

## 11.3 Companion planting

**Figure 11.6**
One companion plant species may naturally produce compounds that repel insects that would otherwise infest another species planted nearby. For example, planting marigolds (*Tagetes* species) in association with tomato plants reduces the number of nematodes, slugs, whitefly and wireworms that infect the tomato plants.

The principle of companion planting is to grow different species of plants together to encourage vigorous and healthy growth and resistance to pests and diseases in all the companion species. Companion planting exploits the plant's natural defences in one of several ways.

'Trap plants' (also known as 'catch plants') can be used to attract problem pests away from the main crop. For example, nasturtiums (*Tropaeolum majus*) attract black fly, thereby keeping them away from adjacent bean crops.

## 'Push–Pull' planting: control of striga weed and stem borers in Africa

Stem borers and the parasitic weed striga (*Striga hermonthica*) are two major pests of maize (*Zea mays*) and other crops in south and eastern Africa. Stem borers can cause a loss of up to 80% of crops, whereas the striga weed can cause a 30–100% loss. A combination of low soil nutrient levels, stem borers and striga can cause the complete loss of a crop. This equates to an annual financial loss of about US $7 billion for maize crops within the affected areas of Africa. Some pesticides are systemic, which means that they are taken up by the plant and transported throughout the plant in the phloem. Other non-systemic pesticides act by direct contact with pests and are not taken up by the plant. Non-systemic pesticides have little effect on stem borers because the pests are protected inside the stem, and so the more expensive systemic pesticides have to be used.

a

b

**Figure 11.7**
The stem borer (a) and striga weed (b) are major pests of cereal crops such as maize.

**Figure 11.8**
Growing the 'push' plant desmodium intercropped between the maize crop and the 'pull' plant Napier grass around the outside of the plot is a management strategy to control the stem borer and striga weed crop pests.

'Push–Pull' is a crop management regime to combat these pests. It was developed by collaboration between the Kenyan-based International Centre of Insect Physiology and Ecology and Rothamsted Research in the UK, based on original work by Jim Miller in the USA. The system requires intercropping with a plant known as silver leaf desmodium (*Desmodium uncinatum*). The desmodium produces a scent which repels the female stem borer and is therefore described as the 'push' plant. At the same time a 'trap plant' such as Napier grass (*Pennisetum purpureum*) is planted around the crop plot. This grass produces compounds that attract the stem borer away from the maize crop and therefore acts as the 'pull' plant. At the same time, desmodium exudes compounds from its roots that hinder the growth of newly germinated striga weed seedlings. An added benefit to using desmodium is that it is a nitrogen-fixing plant and therefore increases the nitrate levels in the soil, improving future crop growth.

## Companion plants as shelters

Companion plants can be employed simply as plant shelters where their presence may shade or provide a physical barrier (intercropping). This can also be achieved by growing hedgerows and shelter belts – a row of trees, shrubs and other plants – alongside a crop. They give the benefit of protecting the crop from wind, frost and temperature variations.

**Figure 11.9**
Two crops are used here, not only because it makes efficient use of the growing area, but also because the maize provides a protective barrier for the crop growing underneath.

Hedges and shelter belts attract insects which are either natural enemies of crop pests or useful for pollination. A shelter belt is less uniform than a hedge as the trees and shrubs are allowed to grow more naturally, rather than being regularly cut.

## Companion plants to control weeds

Companion plants can also produce compounds which can inhibit the growth, development or seed germination of other plants (weeds) within their root zone. This is a mechanism to prevent competition with other plants, known as 'allelopathy'. Within a controlled environment, the use of this kind of companion planting has proved effective but there is doubt about whether it has a real effect in the field, where the effects of competition for light, water and nutrients are more complex. Other concerns are that the effects do not tend to be limited to weed species, and the allelopathic companion plant may also have detrimental effects on the plant species itself if planted continually. It is therefore important to leave enough years between plantings. A plant species which produces allelopathic compounds that actually end up being toxic to the plant itself is called an autotoxic plant.

**Figure 11.10**
Alfalfa (*Medicago sativa*) is a particularly good example of an autotoxic plant. It produces compounds that help prevent other plants from invading, but if grown continuously on the same soil, it will start to be poisoned itself.

Another area of weed control that has been investigated is the creation of synthetic herbicides based on plant compounds that have been shown to have allelopathic effects. An example is the creation of the herbicide Callisto®, which is both a foliar herbicide (acting on the leaves) and a residual herbicide (remaining active within the soil). The active ingredient of this herbicide is mesotrione which is a synthetic chemical similar to a compound extracted from the lemon or scarlet bottlebrush (*Callistemon citrinus*).

**Figure 11.11**
The scarlet bottlebrush (*Callistemon citrinus*) produces the allelopathic compound leptospermone, which was the inspiration for the broadleaf herbicide Callisto®.

a

b

**Figure 11.12**

(a) The tree of heaven (*Ailanthus altissima*) produces the compound ailanthone.

(b) The black walnut (*Juglans nigra*) produces the compound juglone. Both chemicals have been shown to reduce shoot growth of nearby plants significantly (and, in addition, ailanthone has been shown to prevent weed seed germination).

## 11.4 Plants' own chemical defence systems

Plants are unable to move to escape from threats and therefore need an internal mechanism to prevent attack from herbivores and disease. When wounded, plants may produce compounds that seal around the wound site and help prevent secondary infections, but they may also produce an array of compounds that prevent further attack. Many of these compounds are volatile and are created and emitted as a result of tissue damage. As you have already seen in the example of glucosinolates (from Chapter 6), some of these chemicals can be distasteful to chewing pests. When being attacked by pests, some plants release volatile signals that can be recognised by neighbouring plants, allowing these plants to initiate the production of various protective chemical compounds in advance of actually being predated themselves.

As well as producing scents for attracting insect pollinators and for seed dispersal, many of the chemicals discussed in Chapter 6, including the volatile monoterpenoids and sesquiterpenoids, are also produced for defence. They can be induced (produced only when a plant is attacked) or can be constitutive (produced all the time) and are often stored as toxins such as tannins, which are astringent and taste bitter. A plant can produce these chemicals in the leaves, flowers, fruits or roots.

**Figure 11.13**

Conifer species, such as the Sitka spruce (*Picea sitchensis*) and Norway spruce (*Picea abies*), are attacked by white pine weevil (*Pissodes strobi*), which is a major pest in North America. The trees minimise the effect of the weevils by producing a resin that contains a whole array of volatile monoterpenoid and sesquiterpenoid chemicals.

Volatiles, such as the acrid-tasting glucosinolates found in the roots of members of the Cruciferae (Brassicaceae) family, can help defend the plant from soil-borne microbes and herbivores. Volatiles can also attract the natural predators of root-consuming pests.

Plant-produced volatiles can have several modes of action and either directly or indirectly affect herbivores. Directly emitted volatiles can influence a herbivore's physiology, by, for example, disrupting its hormone levels and life cycle. Phytosterols mimic the insect moulting hormone ecdysone and insects that eat plants that emit these volatiles cannot complete their life cycle. Sheep feeding on Australian clover (*Trifolium subterraneum*) show reduced sexual performance and lower levels of lambing. This is due to isoflavone, a type of phytosteroid which is the white pigment in the clover flowers, and which can mimic the female mammalian hormone, oestrogen. Ewes that annually fail to lamb are referred to as suffering from 'clover disease'.

Indirect effects of plant volatiles occur when the plant defence chemicals attract natural predators such as parasitic wasps, mites or flies which will then in turn control the numbers of the pest. This is a complex relationship between the plant, herbivore and natural enemy of the herbivore.

A plant under attack, for instance a mechanical attack or herbivore-induced wounding, can produce 'alarm signals'. These pheromones induce the production of other natural defence compounds, both by all parts of the plant producing them and also by nearby plants. Such signals include a number of gaseous compounds like the plant hormones (growth regulators) ethylene and methyl jasmonate, as well as the compound methyl salicylate (a compound that is very similar to the basis of the drug aspirin).

a

b

c

**Figure 11.14**
When lima beans (*Phaseolus lunatus*) (a) are attacked by spider mites (*Tetranychus urticae*) (b) a whole array of volatile defence compounds are emitted, some of which have been shown to attract the natural enemy of the pest, the carnivorous mites (*Phytoseiulus persimilis*) (c).

a

b

**Figure 11.15**

Conifer species such as the giant redwood (*Sequoiadendron giganteum*) (a) increase production of ethylene and methyl jasmonate when under attack from conifer stem pests. This then induces the trees to produce a whole array of defence responses including wound-resistant resins (b), volatile terpenes and phenolic compounds.

## Biofumigation

Biofumigation is the incorporation into the soil of 'green manure' made from plants that naturally produce volatile chemicals to combat soil-borne weeds and fungal pathogens. One area of research is to find an effective, natural alternative to methyl bromide – a very toxic, ozone-depleting fungicide whose use has been phased out since 2005. Fungicides are chemicals that kill fungi present either on plants or in the soil. Plants from the Cruciferae (Brassicaceae) family, such as bastard cabbage (*Rapistrum rugosum*), rocket (*Eruca sativa*) and mustard greens (*Brassica juncea*), are of particular interest because they contain glucosinolates, which break down in the soil to form the active chemical isothiocyanate, which acts as a biofumigant. Studies into using plants such as bastard cabbage showed a reduction in the fungal pathogens *Pythium* species and *Rhizoctonia solani*. The use of biofumigation is of great interest in organic agriculture but care must be exercised, as any potential biofumigant may be as toxic or as long-lived as the synthetic chemicals they are meant to replace.

As we move forward into the twenty-first century the pressure to maintain crop yields will inevitably increase, and natural plant protection will undoubtedly have a role to play in developing countries and in the developed world, where some consumers and scientists are concerned about the potential dangers of some modern technologies including pesticides, herbicides and genetic modification.

# 12 The impact of humankind on the planet

The greatest and most rapid human-induced changes to the Earth's environment have occurred within the past 50 years. These changes have resulted in improved supplies of food, fresh water, timber, fibre and fuel, which have benefitted the majority of people. However, the rate and extent of these changes are producing unintended and damaging consequences for us all.

Throughout our history, humans have affected the diversity of life on Earth. We, *Homo sapiens*, have altered the environment in numerous ways. Some changes have been radical, often with a negative impact on the ecosystems involved. Some of these impacts have resulted in the irreversible loss of biodiversity. Since 1600, 584 plant species have become extinct. This is important because once a plant species becomes extinct we have lost it forever, and we also lose the interactions that the species has with others around it. As we have seen, plants play central roles in providing humans with food, shelter and medicines, and in maintaining a balanced atmosphere, healthy soils and fresh water. If a plant species becomes extinct before its characteristics have been fully determined, any benefit it may have conferred upon its environment and to humankind is also lost.

Biodiversity is a term derived from **bio**logy and **diversity** and is the 'variation of life at all levels of biological organisation'. Another more specific definition used by ecologists is the 'totality of genes, species and ecosystems

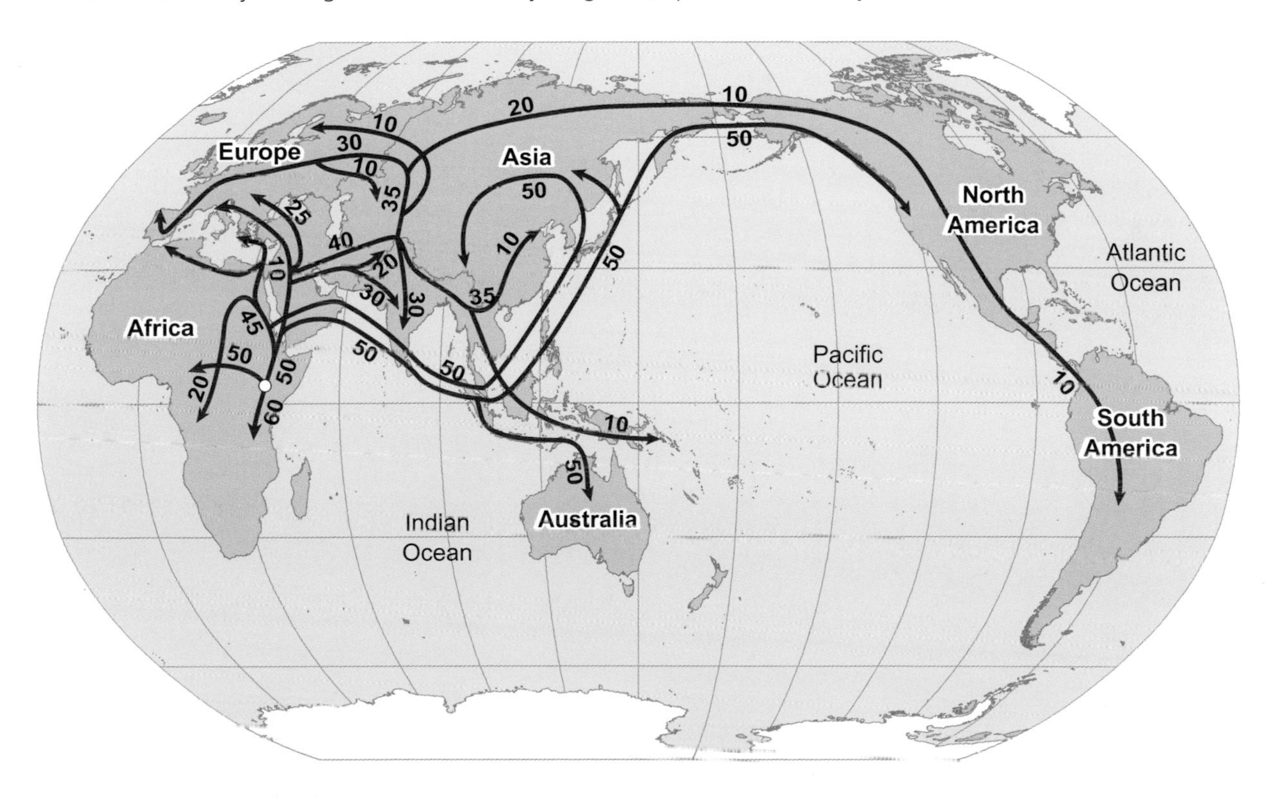


**Figure 12.1**
The patterns of human migration can be traced back to approximately 100,000 years ago when our original ancestors began to move out from Africa. Numbers are thousands of years before present and show when humans reached particular areas. Most of the populations from which we have arisen were in place around 10,000 years ago, although some more remote islands, such as New Zealand, have only been settled by humans in the past 1,000 years.

of a region'. Human exploitation of the Earth's resources through agriculture, forestry, mining, building, and so on has unintentionally put biodiversity at risk by changing the local ecosystems. Over the past 50 years, people have become more aware of their environment and now seek to understand the impact of their actions on biodiversity and to mitigate them where possible.

We humans are one of the major threats to biodiversity. The pressure to ensure that enough food is produced for all the world's inhabitants has resulted in people maximising the area of land available for growing crops. In doing this, forests are cleared, marshland is drained and crops are planted on more marginal land. The plants, animals and fungi that have evolved over thousands of years to live in these habitats are displaced or become extinct and this can have far-reaching and' as yet unrealised consequences. As a species, *Homo sapiens* have proved extremely adept at utilising different environments for their own purposes. Chapter 2 described how our forebears lived a simple hunter-gatherer lifestyle and this served humans well from the point at which they evolved in Africa through to the point at which they colonised other lands.

As humans have moved more easily and frequently within and between continents we have, whether by accident or design, introduced species into ecosystems in which they did not evolve. Some, freed from the restraint of their natural habitats, have become invasive, severely altering the resources that the changed ecosystem can provide.

## 12.1 Loss of biodiversity and plant extinctions

The five main threats to biodiversity are over-exploitation, habitat transformation, invasive alien species, climate change and nitrogen and phosphate pollution. Various different terms may be used by conservation organisations to express these threats. At its worst, loss of biodiversity can push an ecosystem beyond a tipping point and into disastrous and irreversible changes that greatly reduce those ecosystem resources used and valued by humans. Ecosystems consist of a web of many different species interacting within a defined area, so changes to an ecosystem can wipe out a number of species within it. Some widespread species will also exist within another ecosystem and so will not become extinct. But there are some cases where

a

b

**Figure 12.2**
(a) The Monterey pine (*Pinus radiata*) is native to only three small areas in California, USA. These trees help form a habitat in which numerous other plants and animals flourish, including Hickman's potentilla (*Potentilla hickmanii*) and (b) a rare species of orchid called Yadon's piperia (*Piperia yadonii*).

species are highly adapted to very specific habitats within an ecosystem. If changes to the ecosystem result in these specific habitats being lost, these highly adapted species will become extinct. This can then precipitate further changes within the ecosystem.

Some species that are restricted to a small number of habitats become threatened by extinction when their habitats are transformed for human use, through, for example, agriculture and logging. Such habitat-restricted species are under particular threat because the genome of each species is made up of a series of genes, and in plants about half the genes only occur as one variant, or allele. The other 50% of the genes have two or more alleles. When the habitats in which a species occurs begin to shrink and the number of populations falls, there is a risk that some of these alleles will disappear before the species is lost. The loss of alleles from a population results in a reduction in the capacity of that species to vary through reproduction, as there are fewer different forms of its genes to recombine. The capacity of a species to vary from parent to offspring is crucial to the ongoing evolution and hence the long-term survival of the species.

It has been predicted that by 2050 around 10–25% of current (2009) plant species will be condemned to extinction as a result of habitat loss since 1970. The expansion of agriculture will remain one of the major drivers of both habitat and biodiversity loss, resulting in over-exploitation and habitat transformation well into the twenty-first century.

## Over-exploitation

Over-exploitation happens when we harvest more plants from the wild than can grow back in the following season. There are examples of this having occurred throughout our history. In present-day subsistence lifestyles, the extent of this over-exploitation can be seen from the increasing amounts of time it takes to collect loads of fuelwood and medicinal plants. In western lifestyles, hobbies such as gardening can be equally damaging.

**Figure 12.3 - far left**
Although recognised as vulnerable to extinction in the wild and protected under CITES (see Chapter 13), between 1997 and 2008 over 8 million bulbs of snowdrop (*Galanthus* species) were collected from the wild for sale to gardeners.

**Figure 12.4 - left**
Adam's mistletoe (*Trilepidea adamsii*) – an extinct plant painted by Sue Wickison for the New Zealand Plant Conservation Network. The semi-parasitic *Trilepidea adamsii* used to grow on trees in the northern part of the North Island of New Zealand. Whilst the main cause of its decline must be deforestation, it was rare and well known to be so. As a consequence, many plant collectors went out of their way to acquire a specimen. It was last seen in 1954.

## Habitat transformation

Habitat transformation is historically the greatest cause of biodiversity loss. As more and more land is converted for agricultural use, so the area available to the wide variety of wild plants we previously used as hunter-gatherers becomes less. As this area shrinks, so does the number of sites (niches) that are suited to each individual species. As more niches are lost, more unique alleles of a particular gene are also lost, resulting in less variation within the species. Finally, as the number of niches shrinks further, the species itself may well become extinct. On small islands, endemic species (species unique to that island) are particularly vulnerable because of their small population size.

**Figure 12.5 - right**

In the nineteenth century, the expansion of the city of Paris destroyed the only known location of *Viola cryana*, which was endemic to that area. This species was first discovered in 1860, it died out in the wild in 1930. The species was then lost from cultivation in 1950. This picture shows a dried botanical specimen.

**Figure 12.6 - far right**

The tree *Sterculia khasiana* was endemic to the Khasi Hills in Meghalaya, India occurring in the subtropical forests at altitudes of 1,000–1,500 m. The extent of its habitat has significantly declined, largely due to forest fires and conversion of land for agriculture. The species has not been collected since 1877. As such there is no known image of the plant and so shown here is a seed pod from the related *Sterculia ceramica*.

## Invasive alien species

Conservationists believe that invasion by alien plant species is second only to habitat transformation as a cause of species extinction and environmental decline. Invasive alien plants are non-native species that have been introduced relatively recently into a country or an area where they encounter little competition because their natural predators, competitors and diseases may be absent. They out-compete the native species for space, light, water and nutrients and so outgrow them, thus changing the natural vegetation of a habitat. There can be many economic consequences of invasive alien plants. For example, they can affect water supplies and spoil crops causing millions

a

b

**Figure 12.7**

In the UK, some of our rarest plants are under direct threat from alien invasive plants. For example, the aquatic plant brown galingale (*Cyperus fuscus*) (a) is being squeezed out by parrot's feather (*Myriophyllum aquaticum*) (b), a native from Central America which has been introduced into garden ponds as an oxygenating plant and has subsequently escaped into the wild.

a

b

**Figure 12.8**

In the UK a particular problem has been caused by the release into the wild of plants imported for the aquarium trade. For example, Australian swamp stonecrop (*Crassula helmsii*) (a) is choking waterways and costing millions of pounds each year to control, as well as threatening rare native plants such as the star fruit (*Damasonium alisma*) (b).

of pounds worth of damage. They are creating a significant impact on biodiversity around the world.

Most non-native plants have spread through human activities, either accidentally or deliberately. Accidental arrivals may have travelled as seeds in consignments of agricultural produce or nursery plants, as intact seedlings in or on timber, in ship's ballast or on the equipment of armies. Deliberate introductions include plants imported for ornamental, horticultural, medicinal or forestry purposes.

Invasive alien plants are a particular problem on islands where, along with habitat transformation, they have been the main cause of species extinctions in the past 20 years. The rate of introduction of new species continues to be extremely high, although only a few will become invasive. For example, in New Zealand, plant introductions alone have occurred at a rate of 11 species per year since European settlement began in 1840. The islands of French Polynesia have 13 species of aliens listed as noxious weeds; all of them were introduced intentionally as ornamentals or for other purposes.

**Figure 12.9**

Prickly pears are native to the Americas. However, in the 1800s they were imported to Australia to provide two valued functions. The first was to be a natural barrier or fence for farmland. The second was to enable the production of cochineal dye, a pigment that is produced by scale insects that feed on *Opuntia* plants. Unfortunately, the prickly pears did not have any predator species in Australia with the result that they spread very quickly and became weeds that were extremely difficult to eradicate. In 1925, in an effort to control the prickly pear onslaught, the Australians introduced from the Americas the cactus moth (*Cactoblastis cactorum*), whose caterpillars are able to feed on prickly pears. This type of biological control proved very successful and the population of prickly pears became more limited and controlled.

## Japanese knotweed in the UK

Japanese knotweed (*Fallopia japonica*) is a native of Japan and northern China. It was introduced into the UK in the middle of the nineteenth century as an ornamental garden plant. An impressive, tall (2–3 m), perennial plant with arching stems and clusters of white flowers, it was quickly taken up by gardeners and planted widely. The dead stems and leaves persist through the winter, and in spring, new shoots grow from a rhizome so that it quickly forms a dense monoculture which crowds out other plants. The stems and leaves decompose slowly and form a deep layer which prevents native seeds from germinating.

**Figure 12.10**
Japanese knotweed (*Fallopia japonica*).

By the beginning of the twentieth century, gardeners were beginning to notice that the rampant growth and persistence of the knotweed was becoming a problem. This led to plants being uprooted from gardens and dumped; but this only exacerbated the problem, as knotweed thrives in any kind of disturbed soil. As an invader and coloniser, Japanese knotweed has few competitors and has many excellent strategies for establishing itself and resisting removal. It is a dioecious species, bearing male and female reproductive parts on separate plants. Within the UK all knotweed plants are female so spread by seed is not possible but vegetative propagation is very efficient. DNA analysis has shown that all the plants in the UK are clones so must have arisen from one single parent plant.

The rhizomes (horizontal underground stems) may extend laterally for up to 7 m and grow down several metres. Removal of a complete rhizome is almost impossible and even a tiny piece of rhizome can sprout and regenerate a colony of plants. Cutting or strimming fresh stems will also result in many fragments, each of which can sprout roots and shoots within a few days if water is available.

### Impact of Japanese knotweed

As well as being a garden pest and impeding and preventing the growth of UK native plants, this invasive alien can affect the built environment with severe economic consequences. The rhizomes can penetrate brickwork, concrete and tarmac and so undermine buildings, flood defences and river banks, and can also damage archaeological sites. The difficulty and cost of its removal has singled out Japanese knotweed for special treatment as a notifiable weed. The Wildlife and Countryside Act 1981 makes it an offence 'to plant or otherwise cause to grow Japanese knotweed in the wild'. The Environmental Protection Act 1990 also places a duty of care on any person disposing of Japanese knotweed to do so at an appropriately authorised landfill site. This caused an extensive problem for the 2012 Olympic organising committee, who found that the site intended for the stadium in Stratford, east London, was heavily contaminated by this weed. The estimated cost of eradicating Japanese knotweed from this one small site in the country is around £70 million!

### Can this invasive alien be conquered?

Eradication is a lengthy and complicated process. It may involve extensive use of a herbicide such as glyphosate, although special care must be taken if there is any possibility of run-off into water. Physical methods of destruction require removal of plants and soil to designated and authorised disposal sites.

In its native home, Japanese knotweed has predators and parasitic fungi but these are not present in the UK. Research is underway, under very strict quarantine conditions, to find a natural biological control to keep Japanese knotweed under control without affecting other species of knotweed and without causing other environmental problems.

If you recognise an area of Japanese knotweed you should inform your local council as they may have an eradication programme for dealing with this notifiable weed.

### Climate change

Observed recent changes in climate, especially higher regional temperatures, have already had significant impacts on biodiversity, causing changes in species distributions and population sizes. It is predicted that global mean surface temperatures will rise by at least 1–2 °C above current (2009) levels by the end of the century, due to the lasting effects of current and future greenhouse gas emissions. These emissions will rise further in the short term. This temperature rise will condemn 20–25% of plant species to extinction, with longer-lived species at greater risk because they often take longer to reproduce.

### Nitrogen and phosphate pollution

Synthetic production of nitrogen fertiliser has been the key driver for the remarkable increase in food production over the past 50 years. The application of phosphate-based chemical fertiliser increased three-fold between 1960 and 1990, at which point the amount used levelled off. Humans now produce more nitrogen-containing fertilisers than all the other forms of nitrogen produced by all natural pathways combined, as covered in the nitrogen cycle section in Chapter 6. Over the past four decades, run-off of nitrogen- and phosphate-containing fertilisers dissolved in surface water has emerged as one of the most important drivers of change in terrestrial, fresh water, and coastal ecosystems. As yet there are no known losses of plant diversity that can be directly attributed to nutrient pollution.

It is expected that wetlands will be further polluted by the increased deposition of reactive nitrogen from the atmosphere, either in dust or rain, and that this pollution will rise 7–10-fold over the next decade, with unknown consequences. Such airborne reactive nitrogen is typically produced from the burning of fossil fuels, particularly coal.

## 12.2 Recording plant extinctions

Although not originally intended for this purpose, plant classification has provided a baseline for the existence of species and their distribution. It is from this baseline that the consequences of environmental change on plant diversity are monitored.

In 1997, after analysis of 30 years of data, the International Union for the Conservation of Nature (IUCN) listed 380 plant species as globally extinct and 371 species as possible extinctions. This number is surprisingly low when compared with the increasing numbers of plant species that are being recognised as endangered or threatened with extinction. In areas where habitat loss is severe, the threatened number of species can be up to 125 times greater than the reported number of actual extinctions. This difference is explained by the lack of data on the length of time plant species can exist in the threatened state before succumbing to extinction. Plant species have a remarkable capacity to persist in very threatening circumstances. Seeds of a wide variety of plant species can survive in the soil and remain dormant for many years, helping these species to survive. In 1991, 53 species were presumed extinct in Western Australia; however, following an extensive exploration programme targeting these lost species, 19 of these have been rediscovered in the wild.

**Figure 12.11**

*Takhtajania perrieri* was described from a specimen collected in 1909 in Madagascar. DNA studies showed this to be the only species of the Winteraceae family to occur naturally in Africa. In 1993, a small stand of this isolated species was rediscovered in the wild.

**Figure 12.12**

Lobelias, which belong to the Campanulaceae family, grow on the slopes of many Hawaiian mountains and many are threatened due to changes in habitat brought about by humans.

Recorded extinctions vary between different habitats. Areas such as the tropics, which are generally considered to contain the greatest number of plant species per unit area, have fewer recorded extinctions. However, they are also the least well known botanically. Thus the low estimates of extinction reflect the lack of data concerning most plant groups in these areas.

Areas that are better known botanically, such as parts of the temperate or the sub-tropical regions, have many recorded extinctions of species: for instance, the USA has 203 extinct species, Australia has 71, India has 60, Mauritius has 53, South Africa also has 53 and Cuba has 23. These figures relate to known extinctions of species occurring over the past 200 years, during which these areas have been subjected to severe human impact. Probably the most accurate extinction rate is obtained by comparing historical records with present-day survey data. This has been possible for the relatively restricted area of the island of Singapore, where results show that since 1819 about 25% of the original plant species have disappeared. A similar study for the area around Auckland, New Zealand found a local extinction rate of 21% over a 114-year period.

By contrast, the total number of species of Mediterranean higher plant species (that is, excluding mosses, liverworts and ferns) presumed extinct is 35 in a flora of about 23,300, giving an extinction rate for higher plants of 0.15%. Such low extinction rates are found in areas where humanity has been active longest and species incompatible with human development are thought to have become extinct before baselines for the relatively unchanged flora were recorded.

Analysis of the distribution of extinctions amongst plant families supports the view that it is relatively random and not related to evolutionary origins. The nine families with the greatest numbers of extinct species are families with large numbers of species: Compositae (Asteraceae) has 44 extinct species, Orchidaceae has 37, Leguminosae (Fabaceae) has 36, Rubiaceae has 35, Labiatae (Lamiaceae) has 34, Campanulaceae has 27, Palmae (Arecaceae) has 27, Gramineae (Poaceae) has 25 and Rosaceae has 24. Additionally, there is some evidence that some plant families are at greater risk. The first extinctions resulting from deforestation in the Sumatra lowlands seem to have occurred among the epiphytes, such as orchids, that grow on the emergent trees characteristic of a primary rainforest. It is often difficult to identify which of the five drivers of extinction, described above, pushed a species over the edge. In most cases, more than one is involved.

## 12.3 Biodiversity hotspots

To be defined as a biodiversity hotspot, a region must meet two criteria: it must have a very high level of natural and unique biodiversity, and it must have already been subject to great habitat change. Specifically, biodiversity hotspots all contain at least 1,500 endemic species or have 0.5% or more of their plant species as endemics and 70% of their primary vegetation has been transformed from what it was originally. Conservation International lists 34 global terrestrial hotspots. Examples include the California Floristic province in the USA, with around 3,500 plant species of which more than 61% are endemics, and New Zealand (the whole country), which is classed as a hot spot as it has 2,300 species of plants with 1,865 (81%) being endemic.

**Figure 12.13** Map showing the location of selected global biodiversity hotspots.

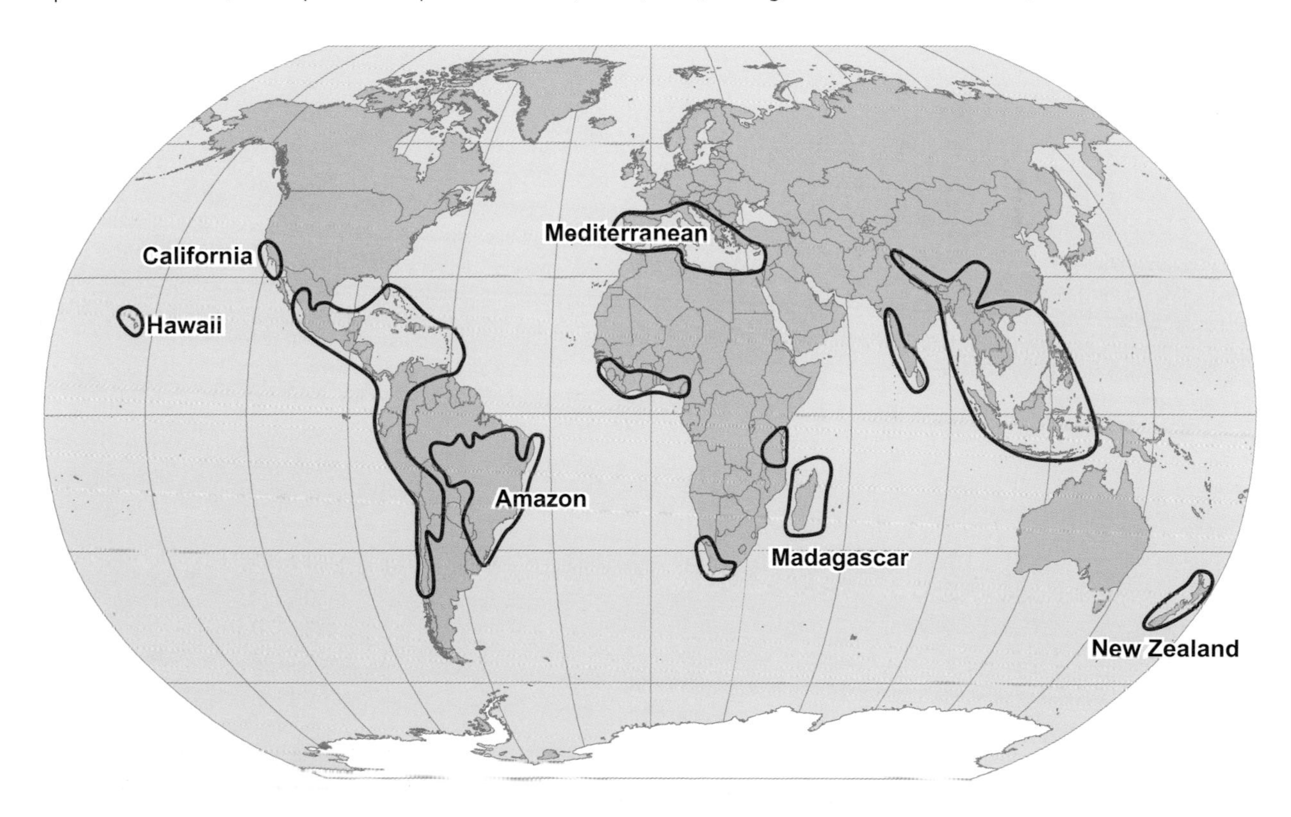

## Madagascar

Figure 12.14

A rainforest habitat on Madagascar.

The island of Madagascar off the east coast of Africa is one of the world's most important biodiversity hotspots. About 90% of its 10,000 species are endemic. The natural vegetation ranges from rainforest to grassland to the unique Spiny Forest of cactus-like succulent plants. Madagascar has a whole continent's worth of plant diversity packed into an island smaller than the US state of Texas. Approximately 80% of Malagasy people rely on subsistence farming for their survival, and the country is losing more than 150,000 hectares of forest a year because of unsustainable farming techniques; this is around one-tenth the area of land covered by greater London. In addition to habitat transformation, there is the threat of over-exploitation of plants, not just for local use but also for the international horticulture trade in the case of plant groups such as orchids, palms and succulents. Plants already in cultivation are continuing to be sourced unsustainably from the wild, putting natural populations at risk. The rarest species cling on as just single populations of a handful of plants. This makes them particularly vulnerable to extinction, as the next tropical storm or forest fire could destroy their habitat and wipe them out completely.

Figure 12.15

Kew's scientists have been working in Madagascar for almost two decades, helping local botanists to catalogue species and identify conservation priorities. Their efforts target species at most risk in three key groups of plants: orchids, palms and succulents. Here, Kew scientists are inspecting a type of Malagasy palm (*Dypsis ambositrae*), a target species for conservation in Madagascar.

**Figure 12.16**
With about 170 species, all but five of them found nowhere else, Madagascar's palms are of global importance. They are also among the most spectacular and horticulturally desirable of all palms. So, like the orchids, they are threatened both by habitat loss and by over-collection.

Many countries and institutions are working with the government and people of Madagascar to conserve its amazing biodiversity in ways that also provide income and improved living conditions.

Orchids form the largest plant family in Madagascar. Kew has developed a method of micropropagation for germinating these orchid seeds so that thousands of seedlings can be produced from a single seed capsule (see Chapter 9). Working in collaboration with the Tsimbazaz Botanical and Zoological Park, seed was collected from endangered species such as *Bulbophyllum elliottii* and plants returned to Madagascar to help support dwindling populations.

Taken together, over-exploitation, habitat transformation, climate change, invasive alien species and both airborne and 'run-off' nitrogen and phosphate pollution have increased the rate of plant species extinction 70-fold compared to the background rate found in the fossil record. Our ingenuity and evolutionary success are destroying the hidden foundations and building blocks from which our livelihoods are built. Recent advances in the accurate recording of plant populations have helped scientists understand where the greatest threats to species occur. It is vitally important that measures to control those factors that cause plant extinctions are found and implemented soon.

# 13 Conservation

The term 'conservation' is generally used to refer to the protection of global biodiversity. The term 'biodiversity' came into widespread use in the 1980s when conservation, and particularly plant conservation, became more of a political topic. Information supplied by organisations such as the International Union for Conservation of Nature (IUCN) showed that more and more species were becoming endangered. The IUCN publish a 'Red List' that details those plants (and animals) whose survival as a species has become threatened. The most threatened species are classified as 'critically endangered', species that are judged to be the next at risk are classed as 'endangered', while species that are in danger but face a lower level of threat are termed 'vulnerable'. It became clear towards the end of the 1980s that many plant (and animal) species were becoming increasingly threatened. The political will to act in order to preserve biodiversity increased, culminating in the United Nations Conference on Environment and Development in 1992. Commonly known as the Rio Earth Summit (after Rio de Janeiro, where the conference took place), it was attended by over 30,000 people from more than 170 countries, including over 100 heads of state. The Summit resulted in the formation of three treaties, sometimes called the 'Rio Conventions': The United Nations Framework Convention on Climate Change (UNCCC), The United Nations Convention to Combat Desertification (UNCCD) and The Convention on Biological Diversity (CBD).

The CBD was the first international treaty to address the threat to the world's biodiversity across all levels. The treaty came into force on 29 December 1993 and by 2009 over 190 countries had ratified it, making the CBD one of the most widely ratified international conventions. The CBD treaty represents a commitment by the nations of the world to:

- conserve biodiversity
- use biological resources sustainably
- share benefits arising from the use of genetic resources fairly and equitably.

Countries (or 'Parties') implement the CBD through their own national laws, regulations, strategies and action plans. The governing body of the CBD is the Conference of the Parties (COP) and it meets every two years to make decisions and review progress. Countries can exchange information on best practices and policies for work in the Earth's major biomes (such as forests and drylands), and also consider issues such as how to control invasive alien species and how to respect traditional knowledge. Each country has a National Focal Point, which can provide information on that country's implementation.

**Figure 13.1**
The Executive Secretaries of the Rio Conventions at the ninth meeting of the Conference of Parties in Bonn, Germany in May 2008.

The CBD recognises that countries have sovereign rights over the biological resources within their borders. It also recognises that, while biodiversity conservation is a common concern for humankind, the cost falls heaviest on countries that are rich in biodiversity but often poor in terms of money and infrastructure. The sharing of benefits in exchange for access to genetic resources is designed to help biodiverse countries reap the benefits from their biological richness and so spread the costs of conservation.

Conservation strategies are enacted in two main ways, either *in situ* or *ex situ*. *In situ* conservation efforts concentrate on retaining a particular habitat and the biodiversity it contains. Those species present can then adapt to the changes that are occurring around them. *Ex situ* conservation refers to removing plant species from their natural habitat and their associated threats, and placing them in a managed human environment, be it a botanic garden or a seed bank. There are potential advantages and disadvantages to both strategies.

## 13.1 *In situ* conservation

*In situ* conservation attempts the holistic approach to conservation, seeking to maintain species, their genetic diversity and interactions in a natural habitat and ecosystem where they can continue to adapt and evolve. Species exist as components of an interactive ecosystem. Plants may rely on animals, birds or insects for pollination and for seed dispersal. Green plants are the producers of the ecosystem, and when consumed transfer nutrients and energy to other levels in the food web. The roots of the plants contribute to the maintenance and retention of soil structure by the stabilisation of mineral particles. The organic material from fallen leaves and dead plants provides food for fungi and bacteria and also nutrients for further plant growth. If the whole ecosystem is protected and conserved, a wide variety of organisms will also thrive and biodiversity will be supported.

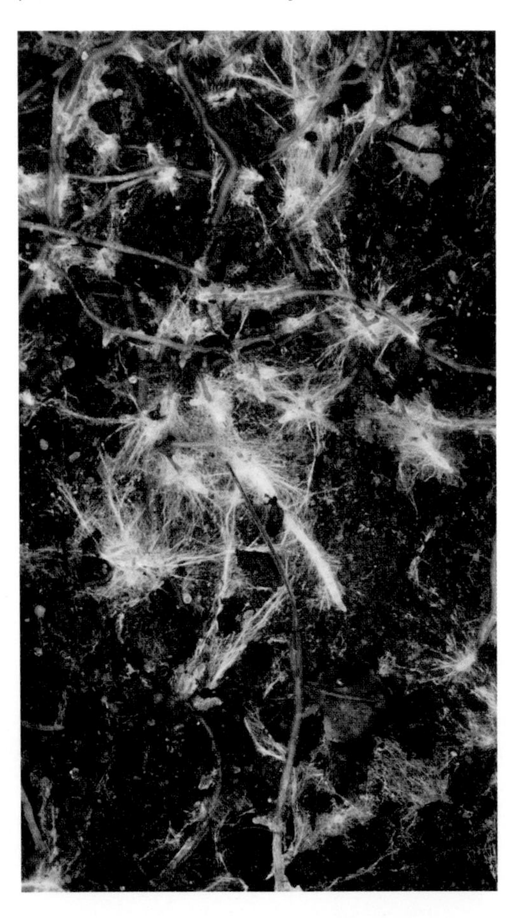

**Figure 13.2**

Fungal mycorrhizae are found associated with most plant species. These mutually beneficial associations promote the growth of both species involved. Sugars made during photosynthesis by the plant are accessible to the network of fungal threads, or mycelium, in and around the roots. The extensive mycelium of the fungus has a large surface area to absorb water, which may be passed to the plant. In times of drought, plants with extensive fungal mycorrhizae have been shown to have increased chances of survival. Plants are also able to obtain mineral ions, such as phosphate, from their associated fungi. *In situ* conservation maintains the fungal connections and thus improves the viability of the threatened plants in the ecosystem.

Competition will occur in an ecosystem and this allows the different species to continue to evolve and possibly to cope with changes in the environment. This is particularly important in times of climate change. Seeds frozen in a seed bank or solitary specimens in a botanic garden do not have the reproductive opportunities that drive adaptation and evolution.

Article 8 of the CBD requires participating countries to establish a system of protected areas or areas where special measures will be taken to conserve biodiversity. The establishment and management of these protected areas is a complex process and can require changes in legislation, work to raise awareness of conservation issues and public education.

To establish whether *in situ* conservation of a particular region is needed, an assessment of the threats facing particular plants is required. In areas of the world where there has been little prior assessment, new technologies such as satellite imagery can be used to estimate the extent of habitat loss. One way of initiating *in situ* conservation is by legislation at global, national and local levels. This should be backed up at a more local level by ecosystem and local community management. Traditionally, *in situ* reserves are established in which the threats of over-exploitation, habitat transformation and invasion of alien species are managed. *In situ* conservation is also encouraged in biodiversity hotspots, where the protection of one single habitat can potentially allow many species within that habitat to be conserved. However, habitat loss – even within biodiversity hotspots – is occurring on a large scale; in some areas 70% of the habitat is already lost. The challenge is clear: given that if only 10% of the habitat is in protected areas, then perhaps as many as 30% of the species will be lost.

*In situ* conservation can occur in many different types of location, on a regional or national scale, and can take different forms in different countries, including: national parks; national and local reserves; wilderness areas; land management agreements; on-farm strategies; community woodlands and grazing lands.

**Figure 13.3**

Yosemite National Park in the USA is around 310,000 hectares (1,200 square miles) and attracts approximately three and half million visitors annually.

## National Parks

The IUCN recognises over 6,000 National Parks and protected areas throughout the world, encompassing around 15% of the land area on the planet. They are categorised as protected areas managed mainly for ecosystem protection and recreation, with restrictions to exclude exploitation or activities that might conflict with their conservation purpose. The opportunities they offer to visitors must be environmentally and culturally compatible with the protected status of the area.

National Parks vary considerably in size but within their boundaries include roads, farms and villages, as well as uncultivated areas. The Rocky Mountain Park (USA) covers an area of 108,000 hectares, Table Mountain Park (South Africa) 25,000 hectares and the Lake District National Park (UK) 23,000 hectares. In the UK, there are 15 National Parks.

## National Nature Reserves

Many countries have established Nature Reserves. For example, in the UK there are 222 National Nature Reserves (NNRs). These sites are smaller than National Parks and generally do not contain buildings or roads. They were established to protect important areas of wildlife habitat and geological interest. In England, NNRs may be found in nearly every county from Lindisfarne in Northumberland to the Lizard peninsula in Cornwall. They are owned and managed by Natural England, the Wildlife Trusts or other approved bodies. The Reserves include a wide range of habitats, including many types of woodland, salt marshes, dunes, heaths, moors and chalk downland, and are valuable areas for study by ecologists. The study of many rare and threatened plants will enable better management of the reserves with improved prospects for the future survival of species.

**Figure 13.4**

Hamford Water in Essex UK is a designated National Nature Reserve and has a specialised set of habitats, including tidal creeks and mudflats. These attract specialised plants that are able to colonise them, such as the very rare sea hog's fennel (*Peucedanum officinale*).

## Wilderness Areas

A Wilderness Area is land that is defined as being biologically intact; that is, it contains all, or most of, the species that it has had through history and is land that is legally protected. It is perhaps an indication of how seriously governments now take conservation that President Obama, in one of his administration's first pieces of legislation since taking office, signed into law the largest conservation bill seen in the USA since 1996. The Land Management Act of 2009 saw 52 new Wilderness Areas formally recognised and protected and added extra area to 26 further Wilderness Areas already protected. In all, this new law added some 81,000 hectares to the land that is classified as Wilderness Area. In addition to this, the bill also created ten new National Heritage Areas, and some of these will aid conservation of plants.

**Figure 13.5**

Mount Baker Wilderness Area in Washington State, USA.

## Land management agreements

In the UK, the Department for Environment, Food and Rural Affairs (Defra) has a programme for an integrated approach to rural development. This includes land management agreements for land next to National Parks or for Sites of Special Scientific Interest (SSSIs). Such agreements may take the form of compensation where intensive farming practices are discouraged in the interests of wildlife conservation. Alternatively, they may be included in a wildlife enhancement agreement for carrying out additional specific management works, such as hedge laying and scrub control.

On-farm conservation measures are designed to maintain or increase the biodiversity on farmlands. Traditionally, fields were home to many relatives of the distinct highly selected cultivars that are grown as crops. These plants contain a range of genetic diversity that has probably never been fully evaluated. Highly bred crop seeds are expensive and a time may come when, due to climate change, disease or pest invasion, these reservoirs of genetic diversity will be needed. If farming practices, such as over-use of herbicides, have eliminated the wild populations, future crop development will be impaired.

In western Europe, there are programmes to educate farmers and to encourage them to adopt practices which are both sustainable and friendly to wildlife conservation without lowering profits excessively. Such measures include:

- leaving stubble in fields over the winter to provide food for small mammals and birds, and to protect the soil from severe weather
- rotating crops with a fallow season, which allows overwintering insects such as butterflies to survive
- planting conservation strips of tough grass and flowering plants across very large fields to provide pollen and nectar for insects
- the creation of wildflower meadows to help maintain plant biodiversity and provide food for beneficial insects
- managing hedges and ditches to provide drainage, windbreaks, shelter and food for wildlife
- providing buffer zones around the edges of fields and near watercourses to prevent fertiliser and pesticide run-off, and to provide wildlife habitats.

Land management agreements in the emerging economies, although not so well legislated and developed, have been able to demonstrate some real benefits for the local populations.

Community involvement in organised management of woodlands and plantations results in increased income to the community and promotes biodiversity. *In situ* conservation is potentially very effective because it seeks to protect whole ecosystems. However, it does not and cannot conserve all threatened species because some threatened plants will be outside the protected areas. Also, protected areas do not always guarantee protection of species. Even where whole ecosystems are protected, they are subject to threats like climate change, invasive alien species, natural disasters and political instability. Therefore, there is a need to complement *in situ* conservation with *ex situ* measures such as seed banks.

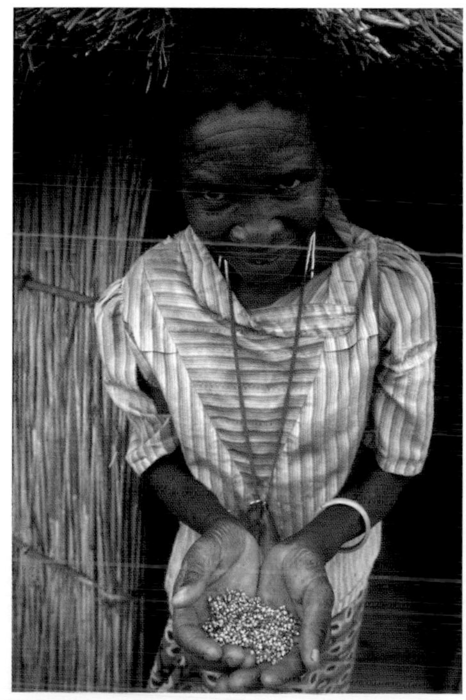

**Figure 13.6**
The UK Darwin Initiative is working with farmers in Africa to promote the conservation of their local crop species and to improve the standard of living for the people. One successful project has taken place at Ipongo in Zambia. It was based on a grassroots-oriented farmer-group approach. Training was given through Farmer Field Schools and the skills and techniques learnt were passed on to others. As well as their normal crops (mainly maize, sorghum, millet and groundnut), the farmers were encouraged to grow two extra crops, including a legume to improve the soil fertility. Seed multiplication and seed fairs using local materials helped to increase crop diversity and to improve the supply of seeds. Through meetings and training, the whole community became involved and enthused.

## 13.2 *Ex situ* conservation

*Ex situ* conservation of threatened species outside their natural habitats provides a complementary strategy against extinction. Currently it is the only approach available to deal with the threats of climate change and nitrogen and phosphate pollution if plant populations are unable to adapt to the changes that they are exposed to. *Ex situ* collections provide a valuable, readily available source of accurately identified plant material that can be propagated and used to help ecosystems in damaged habitats to recover. Such collections are also a source of plant material for investigation into novel properties for human well-being. In this way, they can also limit the collection of more plants from the wild.

The priority categories of plant groups (families) for *ex situ* conservation are those that are:

- the wild relatives of crop plants
- of special scientific interest, such as endemic species
- in danger of extinction at global, national or local level due to one or more drivers of biodiversity loss (see Chapter 12)
- of local economic importance, as these are the plant groups that are likely to be over-exploited: particularly those used by subsistence farmers as food crops, medicinal plants or dyes, or for local industry, horticulture or crafts
- typical plants of an area, which may be needed in future restoration projects, such as after mineral extraction
- local 'flagship' species, which will catch the imagination of people and stimulate awareness in conservation.

The importance of *ex situ* conservation is demonstrated in Article 9 of the CBD, which states that 'Each Contracting Party shall adopt measures for the *ex situ* conservation of components of biological diversity, preferably in the country of origin'.

One of the CBD programmes is the Global Strategy for Plant Conservation (GSPC). The GSPC translates the aims of the CBD into practical measures. One of the targets requires that '60% of threatened plant species is accessible in *ex situ* collections, preferably in the country of origin, and 10% of them included in recovery and restoration programmes'. The focus of this target is on higher seed-bearing plants (conifers and flowering plants), Pteridophytes (ferns) and Bryophytes (mosses), which are being held in collections as a safeguard against danger or loss. The GSPC encourages the holding of such collections in more than one place in the same or another country as a further protective policy against hostile conditions, disease or civil unrest.

The three main types of *ex situ* conservation are:

- botanic gardens, which are living, growing collections of a wide variety of plants
- seed banks, where collections of seeds are held in a state of quiescence
- field gene banks, which are living collections of a particular genus.

### Botanic gardens

Throughout the world there are over 1,300 botanic gardens dedicated to conserving living collections of plants. These gardens vary considerably in their size and in their history. Botanic gardens are distinguished from other beautifully landscaped collections of plants by including plant study and research among their principal aims. Many botanic gardens started out as 'simple' gardens: collections of plants used for their medicinal properties, often attached to monasteries or universities.

Britain's oldest botanic garden is in Oxford. It was established in 1633 for the teaching of botany. By the end of the seventeenth century, the curator Jacob Bobart had drawn up a catalogue of seeds collected from the garden and was exchanging seeds and information with other gardens. The exploration and colonisation of other parts of the world led to the importation of many exotic species for their novelty or their economic or medicinal importance, and the content of living botanic collections changed to include non-native plants. Plant collectors from Kew, such as Joseph Banks, were setting up botanic gardens in newly acquired territories as early as 1780.

**Figure 13.7**

The botanic garden of Padua in Italy, founded in 1545, is regarded as the oldest university garden. Originally it grew medicinal plants as teaching aids for the university, thus establishing the link between botanic gardens and their use for scientific study and education.

**Figure 13.8**

The Chelsea Physic Garden in London was set up in 1673 as the Apothecaries' Garden to train apprentices in the recognition and uses of plants as medicines.

This heritage has provided botanic gardens with the experience and knowledge to play a significant part in *ex situ* conservation. Botanic Gardens Conservation International (BGCI) has a global membership that works on an international scale to safeguard tens of thousands of plant species from extinction. More than 400 botanic gardens around the world contain areas

**Figure 13.9**

The Herbarium at the Royal Botanic Gardens, Kew was established over 150 years ago and in that time has amassed over 7 million different specimens. The collection, particularly in its early days, grew as a result of bringing together individual private collections. Many well-known scientists and explorers have donated their personal collections, including Sir William Hooker, Charles Darwin, Joseph Hooker and David Livingstone. The herbarium specimens are still used by research scientists to identify accurately plant specimens and, in consequence, the collection is still expanding with around 30,000 new specimens added every year.

designated for conservation and also look after similar areas outside their gardens. These gardens provide plant material for research and propagation. Many botanic gardens have herbaria, which are collections of pressed dried plants that provide references for plant identification and for advances in the study of taxonomy and classification. Many herbaria contribute to the establishment of databases giving the historic and present distribution of plants.

## Seed banks

Seed banks provide an insurance policy against the extinction of plants in the wild and provide options for their future use. They complement *in situ* conservation methods, which conserve plants and animals directly in the wild.

**Figure 13.10**
A seed bank showing the collection of dried seeds housed in small sealed glass jars at sub-zero temperatures.

*Ex situ* conservation of seeds has some useful advantages. Seed banks represent an efficient and cost-effective way of conserving a wide range of the genetic variation of individual plant species; seeds can be collected from natural populations without risking their continued survival; seeds occupy little space, and they require relatively little attention. Properly stored, most species retain high levels of viability for decades, if not centuries. In addition, due to the portable nature of seeds, their storage can easily be duplicated in different places, thereby limiting their vulnerability.

Seed banks provide options for the future conservation and utilisation of plants. Even if a species is lost in the wild, if its seed is stored, it should always be possible to germinate it and so reintroduce these plants. Seed banks can provide a source of high quality and genetically diverse material for the potential recovery and restoration of threatened species and ecosystems.

Effective germination and propagation protocols are developed for species conserved in seed banks to ensure that the seeds contained within them can be germinated if required. Research into seed storage behaviour maximises our ability to use seed banks, and can enable the sustainable use of species that might otherwise be at risk. For example, as was shown in Chapter 7, the bark of the African cherry (*Prunus africana*) can be used to treat BPH, but in Burkina Faso the tree is becoming endangered due to over-harvesting. Staff and partners of the Millennium Seed Bank Project at Kew have developed a method for germinating its seeds, which could lead to the establishment of African cherry plantations, counteracting the over-harvesting of the wild plants.

**Figure 13.11**
An example of the variety of seed shapes that can be present within a seed bank.

## How does seed banking work?

Most species produce seeds that are tolerant of desiccation and when dried and stored at sub-zero temperatures in hermetically-sealed low-moisture containers will remain viable for considerable periods. After collection, seeds are checked for parasites, insect damage and general health and then dried to 3–7% moisture content, depending on their oil content. (Typically seeds contain around 14–20% moisture before drying.) As a general rule, reducing the moisture content of a seed by 1% will double its storage viability. To maximise seed longevity, most seed banks dry the seeds using fairly cool conditions (15–18 °C) and low humidity (15% relative humidity).

Eventually, during drying, equilibrium is achieved with no net movement of moisture either into or out of the seed; usually this takes less than a month. Once dried, the collections are packaged in a variety of different airtight containers prior to storage at –20 °C. This is the economic optimum; cooling to lower temperatures will increase longevity further but the costs increase disproportionately. Through the combination of drying and cooling, lifespans of decades, centuries and, for some seeds, millennia can be achieved.

At the start of storage, and every so often subsequently, some seeds from each collection are taken out of the bank and germinated, to make sure that they are still viable. Many seeds naturally undergo a period of dormancy before germination. In the northern temperate regions, seeds are dispersed in autumn and remain dormant in the soil over the winter and then germinate in the spring. During dormancy, the living embryo inside the seed is in a state of suspended animation. Breaking dormancy may involve several factors including temperature, moisture, the amount of light the seed receives and the presence of any plant growth regulators.

### Field gene banks

A field gene bank arises from the collection and subsequent planting of a number of different varieties or cultivars of species such as those that produce few or no seeds, those whose genetic variation is most easily maintained through clonal propagation or those that have a juvenile phase. Field gene banks are often used for plants such as cocoa, rubber, coffee and various types of grasses.

## 13.3 Conservation of island habitats

Islands are a specific type of landmass that warrants particular conservation interventions. Islands can be home to a diverse range of habitats, from sub-Antarctic heath to cactus scrub and tropical forest, and as a result of their isolation, islands tend to have a high number of endemic species. However, many island floras are under great threat from development pressures leading to habitat transformation and from invasive alien species.

**Figure 13.12**
The remote South Atlantic island of St Helena has lost six endemic species of plant and seven endemic birds since the island was discovered in 1502. These species were almost certainly driven to extinction as a result of invasive species. Portuguese explorers introduced goats as a source of fresh meat and rats came with them as unwelcome guests on their ships. These were followed by many invasive plant species that overwhelmed native floras and filled the gaps created by goat over-grazing. Gradually a unique island ecosystem was transformed into one dominated by invasive alien species with the endemic flora surviving in inaccessible cliffs and high peaks. Today, conservationists are fighting to save yet another species on the brink, the bastard gumwood, *Commidendrum rotundifolium*, currently known from only one pure individual.

Preventing alien species from arriving on islands is the first and most cost-effective option to preserve the biodiversity of islands. There is much emphasis on biosecurity, which is ensuring that plants and animals are not brought onto the islands either unintentionally or even intentionally. However, many small islands lack adequate biosecurity and continue to suffer from new introductions.

## Alien invasive species in the Caribbean islands

In 2005, Kew conservation staff discovered an invasive scale insect in the Turks and Caicos Islands (TCI). The pine tortoise scale, *Toumeyella parvicornis*, feeds on the native pine tree, *Pinus caribaea* var. *bahamensis*. This was the first time this insect has been detected in the Caribbean, and it was a concern locally as the pine is the national tree and an endemic variety of Caribbean pine found only in this region. The scale insect is a forestry pest in North America, and was probably introduced on live Christmas trees imported into TCI without adequate quarantine. It sucks the sap of pine trees, causing greatest harm to seedlings and young saplings, and it can also overwhelm and kill mature trees. The scale insect is spreading within the tropical and subtropical coniferous forest regions of the TCI and many trees are dying. As a consequence, the TCI are reviewing their biosecurity measures.

Until effective biosecurity is in place, island authorities have to try and deal with invasions. Early detection, such as was the case with the pine tortoise scale, will provide more options to deal with the invasion. Eradication may be possible. A single large patch of the highly invasive legume vine kudzu (*Pueraria lobata*) was identified on Bermuda with the help of Kew legume specialists. This patch was completely destroyed, hopefully eradicating kudzu from Bermuda, but it is essential to keep monitoring for possible re-occurrence and a system to do this has been introduced.

On the island of Anegada in the British Virgin Islands an alien coastal shrub, *Scaevola taccada*, has been identified. This species is known to be a major invasive alien species in several Caribbean islands, including Grand Cayman where it has out-competed the native inkberry (*Scaevola plumieri*), which has been reduced to two populations of fewer than ten individual plants. The locations of this species around Anegada have been mapped and complete eradication is still possible. However, if much more time goes by then the only realistic option will be to try and control its spread.

**Figure 13.13**

Originally introduced into Florida as a beach stabilising and landscaping plant, *Scaevola taccada* is gradually spreading throughout the Caribbean turning species-rich coastal vegetation into a monoculture.

**Figure 13.14**

Once an invasive species is well established, control is the only real option and this is a very resource-intensive activity. The Australian pine, *Casuarina equisetifolia*, is a coastal tree native to Australia and the Pacific that has been widely planted as a shade tree throughout the Caribbean. It self seeds easily and has become a major threat to native coastal vegetation and disrupts sea turtle nesting. Mechanical removal of seedlings and saplings is being used to control spread in many islands, whilst the use of systemic herbicide on the cut stumps of larger trees is another effective control.

There is increasing interest in being able to conserve the plant genetic diversity present on Earth. The best way of doing this is by preserving large areas of land in which plant species are able to remain within their natural habitats. However, this is not always possible as the areas required are unlikely to be made available, and so various types of seed banks that can help to maintain the genetic diversity of plant species may be the next best answer.

# 14 Plant collecting and trading

Humans have transferred plants from one location to another ever since we adopted a settled lifestyle. The Romans, in particular, introduced edible and medicinal plants of Mediterranean origin to all areas of their empire. Originally, the transfer of plant material was related to plants that could be used in a particularly beneficial manner. Later, collecting plant material for its aesthetic horticultural value became a goal. The great age of plant collecting began in the seventeenth century and lasted until early in the twentieth century. Not surprisingly, several of these collectors were British. During this period, Britain had a large powerful Navy and was in the process of establishing an empire across the globe. Expeditions were mounted to find new territories, and explorers and plant collectors were often present on such trips. The type of plant material they collected varied, and although it sometimes had agricultural or other commercial importance, frequently it was collected more for its horticultural value.

## 14.1 Early plant collectors

### The Tradescants

John Tradescant the younger (1608–62) and his father, John the elder (1570–1638), were early plant collectors. Between them they worked for some of the highest-ranking noblemen and royalty in England, including Robert Cecil, Duke of Buckingham and both Elizabeth I and Charles I. They supervised some of the great gardens of that period and also travelled extensively collecting plant material as they went, and bringing it back to England. Probably the best-known plant they introduced was the horse chestnut (*Aesculus hippocastanum*). As well as collecting plants, they also imported other collectables from their expeditions, which became the basis of what is now known as the Ashmolean Museum in Oxford, UK.

The Tradescants were examples of plant collectors whose work was funded by wealthy beneficiaries. As time went on, this method of funding collecting trips continued but there was an increase in trips by wealthy individual gentlemen who financed themselves for their own purposes. Such a man was Joseph Banks, who in time was to become the *de facto* head of what is now known as the Royal Botanic Gardens, Kew.

a

b

**Figure 14.1**

(a) John Tradescant the elder.

(b) The genus *Tradescantia* (spiderwort) was named after the elder John Tradescant although some people believed the genus was named after his son.

## Sir Joseph Banks

Sir Joseph Banks was a keen plant collector and he effectively took charge of Kew Gardens from 1773. He instigated plant-collecting expeditions, encouraged the transfer of crop plants around the world and supported the development of botanic gardens in countries other than Britain. It was Banks's enthusiasm and entrepreneurial skills that encouraged other plant collectors to discover new species and bring them to Kew. His vision helped greatly to increase the number of recorded plant species now known. As the richest man in Britain, Banks had very influential contacts with both the East India Company (see Chapter 8) and also with royalty. It was Banks who organised the infamous HMS *Bounty* trip, headed by Captain Bligh, to the South Pacific to collect breadfruit trees, the fruit of which were to be used to feed slaves in the West Indies. One of the many important species that Banks brought back from his travels was the sacred lotus (*Nelumbo nucifera*).

a

b

**Figure 14.2**

(a) The Australian genus *Banksia* that Linnaeus the younger named after Banks as an acknowledgement of his significant contribution to the world of plant collecting.

(b) Sir Joseph Banks (1743–1820).

## Sir Joseph Dalton Hooker

Sir Joseph Dalton Hooker trained as a doctor in Edinburgh, but his principal interest was in botany. He travelled widely in search of plants; for example, between 1839 and 1843, he was assistant surgeon and botanist on HMS *Erebus*, which was sent to Antarctica and visited a number of locations including Australia and New Zealand *en route*. Hooker was interested in the classification and distribution of plants from different parts of the world. His expertise was recognised by Darwin who asked him to classify the plants he had collected in the Galapagos Islands. Hooker and Darwin remained firm friends. Indeed Darwin had a portrait of Hooker over the fireplace in his study at Down House. Hooker also remained one of Darwin's strongest allies when he met initial resistance after publishing his theory on natural selection.

Hooker journeyed through northern India and Nepal (1848–51), surveying the flora there and sending specimens back to Kew. Among them were many previously unknown species of rhododendron, some of which can still be seen in the Rhododendron Dell at Kew. He became Director of Kew in 1865, succeeding his father in the role, and remained in post for 20 years, during which he continued his enthusiasm for global plant collecting.

a

b

Figure 14.3

(a) Sir Joseph Dalton Hooker (1817–1911).

(b) *Crinodendron hookerianum* was named in his honour.

The dedication of plant collectors like Banks and Hooker helped to establish the great botanical collections of the eighteenth and nineteenth centuries. These botanic gardens were not created simply for the pleasure of the masses; they also played a role in enhancing trade opportunities. Botanic gardens became the testing places for new plant introductions and they came to function as part of a network that spanned the world, facilitating plant-exchange and trade. The best example of this is the help that Hooker and Kew gave in 1859–60, which enabled fever bark trees (*Cinchona*, Figure 7.3a) to be exported from the Americas to British-controlled India. As shown in Chapter 7, these trees were economically very important as the drug quinine could be extracted from the bark. As Britain had many colonies in the tropics, the ability to treat malaria effectively was of prime importance.

Another example in which Hooker assisted in a trade opportunity was in the 1870s when rubber trees (*Hevea brasiliensis*) were exported from Brazil to be grown in the British colonies in the Far East, notably Malaysia and Singapore. The sap from these trees can be tapped (extracted) and the latex it contains can be used for a whole range of industrial purposes from waterproofing to the production of condoms.

Figure 14.4

A rubber tree (*Hevea brasiliensis*) being tapped to collect the milky latex liquid from which rubber is produced.

As the twentieth century unfolded, many plant products from across the world were being traded, and it was clear that some form of international law or set of rules was required if certain species of plants were to be protected from extinction in the wild. Obviously such laws needed to have co-operation from as many countries as possible. The answer came in a United Nations Convention.

## 14.2 Convention on International Trade in Endangered Species

Established in 1973, the Convention on International Trade in Endangered Species (CITES) of wild flora and fauna was the first United Nations environmental convention. It aims to regulate and monitor the international trade in plants and animals threatened, or potentially threatened, in the wild by such trade. CITES regulates trade in listed plants and animals by means of an internationally accepted permit system. CITES allows trade in plant species that can withstand current rates of exploitation, but prevents the trade of species facing extinction. Over 170 countries have ratified the Convention, and this commitment by governments, translated into national laws, makes CITES an extremely powerful weapon in the battle for sustainability and maintenance of biodiversity against unacceptable exploitation for international trade.

**Figure 14.5**

The star cactus, *Astrophytum asterias,* is listed in CITES Appendix I. All cacti and orchids not included in CITES Appendix I are included in Appendix II.

At the core of the Convention are three lists or appendices of increasingly threatened species. There are over 25,000 plant species subject to CITES controls: around five times as many plants as animals.

Appendix I of the Convention lists more than 300 plant species; these are the species most seriously threatened with extinction due to international trade. No commercial trade of these plants is permitted. Permits can be granted only for genuine research projects concerned with the biology or conservation of a particular species. Trade is allowed for hybrids of Appendix I species or for plants that have been artificially propagated. CITES defines artificially propagated plants as those that are grown under controlled conditions from seeds, cuttings, divisions, callus tissues (soft tissue that forms over a wounded or cut plant surface) or other plant tissues.

**Figure 14.6**

*Protea odorata* is an example of a plant included in CITES Appendix II.

Appendix II contains more than 28,000 species. These are species that at present are not threatened with extinction because of international trade, but may become so if unregulated trade continues. Plants in this category may be traded if it can be demonstrated that removing the plants from the wild will not push the species closer to the brink of extinction. Trade of both wild and artificially propagated material is allowed under licence. Seeds, pollen, seedlings grown *in vitro* or tissue derived from micropropagation, and cut flowers from artificially propagated plants, are exempt. If they are CITES registered, scientific institutions may receive and exchange living plants, seeds, pollen, DNA or herbarium specimens for research purposes. This facility is essential for the processes of *ex situ* conservation carried out by seed banks and botanic gardens.

Appendix III contains just ten plant species. These are native plants that are threatened with extinction locally through commercial exploitation. These species may grow elsewhere in the region so special international co-operation is needed to conserve the plants in their particular locations and to prevent over-exploitation.

## 14.3 Over-exploitation due to trade

### Trade for ornamental use

In the developed world there is a huge consumer demand for plants for ornamental and horticultural use in glasshouses, gardens and parks. Their removal from the wild has an important effect on species survival rates and can cause significant damage to the environments and ecosystems from which they are removed.

**Figure 14.7**

(a) Asian slipper orchids are highly desirable and collectable and are vulnerable to over-collection from the wild. Many grow in small colonies that can be seriously affected by the removal of even a few individuals. *Paphiopedilum rothschildianum*, shown here, has been reduced to just a handful of plants.

(b) Hybrids of orchids, such as this *Phalaenopsis*, are far more common in cultivation and trade than the species themselves. Most hybrids do not occur naturally in the wild but they look very much like the wild species, so they are included in CITES lists to protect the wild species.

**Figure 14.8 - right**

Cacti are endemic to the Americas with the exception of just one genus, *Rhipsalis*. Species range from the minute dwarf cacti hidden in the sands and gravel of the desert to this giant Saguaro cactus, that well-known backdrop to every cowboy movie. The first reports of cultivation of cacti in Europe date back to the 1500s. Today, Europe's horticultural industry propagates millions of cacti each year. However, there remains a persistent demand for mature specimens of species collected from the wild.

**Figure 14.9 - below**

The cycad *Encephalartos woodii*, which grows in the Temperate House at the Royal Botanic Gardens, Kew. This is the rarest plant in Kew Gardens, and is extinct in the wild. Cycads are amongst the most primitive of plants and are widely cultivated as ornamentals by gardeners in mild climates, and in Europe as large decorative container plants. Many are sought after by specialist collectors and the removal of whole plants and collection of seed from the wild are having a serious impact on populations of these plants. The poaching of large plants from the wild remains a serious problem wherever cycads occur.

**Figure 14.10 - below**

There are 21 species of cyclamen, native to Europe, western Asia and North Africa. They are of great horticultural interest, with many different colour forms and leaf markings, and four species are widely grown in gardens. *Cyclamen persicum* is commonly available, and artificially propagated. Cultivar plants are now exempt from CITES control. Species plants and tubers still require permits and are CITES listed. In Turkey, sustainable collection from the wild provides employment and income for local people.

**Figure 14.11 - right**

Tree ferns are distributed in the Americas, South-East Asia, Australia, New Zealand and Africa. The image used here was taken on an expedition in Malawi. Species from three genera, *Cyathea, Cibotium* and *Dicksonia*, are controlled by CITES. *Cyathea* and *Dicksonia* are traded as growing plants, but more usually as sawn-off trunks or trunk sections. Trunks are also traded as blocks and pots, which are often used in the horticultural trade to grow other plants, especially orchids. Such products are also subject to CITES controls. *Cibotium* is traded as dried roots and also as an ingredient in traditional Chinese medicine. The *Cibotium* traded to date is only known to come from the wild, with no known large-scale propagation for the medicinal trade.

## Carnivorous plants

The white-topped pitcher-plant, *Sarracenia leucophylla*, is a member of a large group, known as carnivorous plants, many of which are included in the CITES Appendices. The genera controlled are *Nepenthes*, *Sarracenia* and *Dionaea*, the last of which only has one species and is well known under its common name, the Venus flytrap. Carnivorous plants are often found growing on poor, nitrogen-deficient soils; they obtain the nitrogen they require by digesting the bodies of their prey, such as flies. In the south-eastern United States, habitat transformation poses a significant threat to *Sarracenia* and *Dionaea*. In addition, the pitchers of certain species are cut and harvested for the floristry trade and there is concern about the sustainability of harvesting from the wild. Pitcher plants are easily propagated artificially from seed or rhizomes. Venus flytrap is easily artificially propagated, but is still collected from the wild.

a

b

**Figure 14.12**
(a) The white-topped pitcher-plant, *Sarracenia leucophylla* and (b) the Venus flytrap, *Dionaea muscipula*.

## Trade for medicinal use

Many thousands of species are used in traditional medicines. In 2007, approximately 500,000 tonnes of medicinal and aromatic plants were traded internationally, with a reported value of US $8 billion according to customs data. Amongst these species, 60 are so heavily exploited that CITES listing is needed to provide them with special protection.

One example is *Hoodia gordonii* (Figure 7.14), which grows in the Central Kalahari and Makgadikgadi National Parks in Botswana and which the local San people have used to allay pangs of thirst on long hunting expeditions. South Africa's Council for Scientific and Industrial Research has isolated a compound called P57 from *Hoodia gordonii*, which has been found to act as an appetite suppressant. Drug companies are now developing this compound into a slimming pill. As the demand for *Hoodia gordonii* has increased, other species of *Hoodia* are under threat from collectors. All species of *Hoodia* are CITES listed.

## Trade for wood products

Over 1,000 tree species are traded internationally but only about 80 trees are CITES listed.

Big-leaved mahogany (*Swietenia macrophylla*; Figure 3.8) and ramin (*Gonystylus* species) are included in Appendix II. Another South American species Spanish cedar (*Cedrela odorata*) is included in Appendix III. CITES controls are limited to logs, sawn wood and veneer sheets. CITES, in effect, controls the tree and its parts and derivatives including manufactured material and scientific specimens.

There is worldwide concern about the levels of logging and trade in timber, mainly centred on the contribution that deforestation makes to increasing carbon emissions. It is to be hoped that more trees will be added to the CITES list and be protected from extinction.

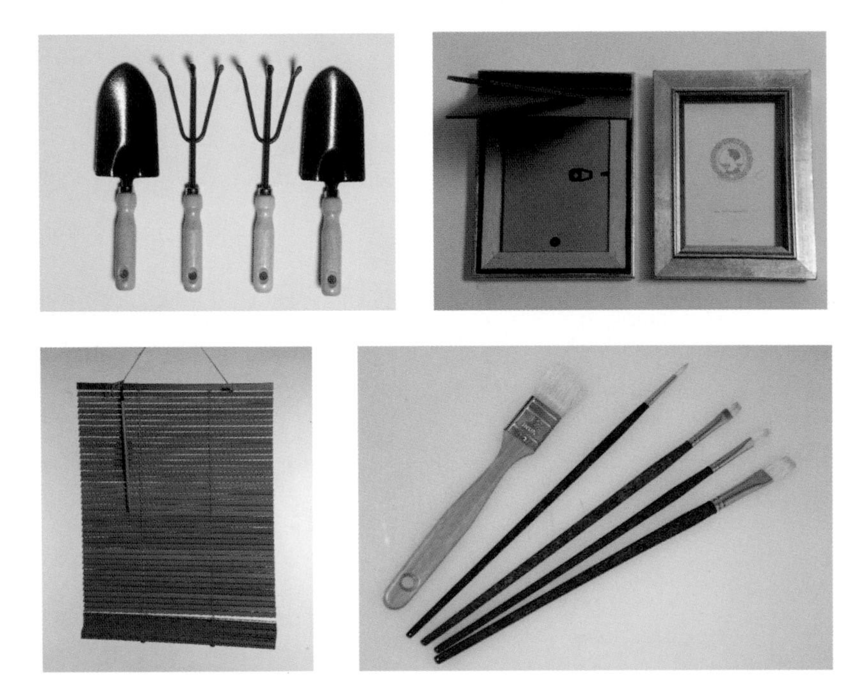

**Figure 14.13** Some articles made from ramin wood.

## Biopiracy

Some people have used the term 'biopiracy' to describe the appropriation of plant genetic material and associated knowledge from developing countries, and the subsequent commercialisation of that material and knowledge without compensation to the owner(s) of that plant or information. It can be argued that biopiracy contributes to inequality between developing countries that are rich in biodiversity, and developed countries with more capacity for commercial research, and has been described as 'colonialism of the modern age'.

In order to address biopiracy, the Convention on Biological Diversity (see Chapter 13) has raised aspirations around 'access and benefit sharing'. The CBD states that genetic resources should be conserved and used sustainably and that benefits should be shared. The benefits may not be purely financial; they could also include community development activity, training opportunities and joint research. Anyone wishing to use or exploit genetic resources must seek prior informed consent for access from providers, mutually agree terms of use and share benefits fairly and equitably (Article 15). These provisions of the CBD are being translated into national legislation, so a country may impose fines and even custodial sentences for illegal collection of material and information. Implementation of these laws against national and multinational companies is not always easy. However, the threat of being labelled a 'biopirate' has led many companies and institutions to develop more stringent procedures and policies – including abandoning projects to characterise genetic materials – so that they are seen to act within the law and maintain their reputation. Many provider countries argue that they need support in enforcing the access and benefit provisions of the CBD, and particularly with compliance and enforcement of national law. The development of an international regime on access and benefit sharing (scheduled to be adopted in some form in 2010) would spread the responsibility between users and providers of genetic resources.

## 14.4 Present-day plant collectors

In contrast to their predecessors, present-day plant collectors now go out into the field to collect seeds from plants with the express intention of preserving the genetic diversity for future generations.

### Michael Way

Michael Way is a modern-day plant collector who is a trained ecologist working on the UK Millennium Seed Bank Project (MSBP). The MSBP aimed to collect and conserve the seeds of 10% of the wild plant species of the world by 2010 (with the target being actually achieved in October 2009), rising to 25% by 2020. As well as conserving seed from all the native UK species, the project is particularly concerned with semi-arid or dryland areas, which make up more than 40% of the Earth's landmass. The seed collected does not come from just any plants; it includes seed from the rarest, most threatened and most useful species known to humanity. Michael co-ordinates seed collecting from North and South America including regions of Chile, USA and Mexico, which are considered to be global biodiversity hotspots (see Chapter 12). He has been involved in the collection and conservation of more than 1,000 species. As well as seeds, specimens of the plants are collected, pressed and dried so that their identification can be checked at the Herbarium at Kew. Data about the location of the plants, their altitude, latitude, longitude, soil conditions and surrounding vegetation is recorded, along with their flowering and fruiting times and their uses to the local people.

**Figure 14.14**
Michael Way collecting *Ephedra andina* in Chile.

Some collecting trips are especially exciting, as was one in Chile in January 2008. Working with colleagues from the Chilean National Agricultural Research Institute (INIA), Michael and his team collected cuttings from *Plazia cheiranthifolia* which had been believed extinct until rediscovered recently, and a large collection of seeds from the endangered bamboo *Chusquea ciliata* in the hills above Valparaiso. Michael said, 'As we looked over the brow of the hill, we could see the pale yellow foliage of the bamboo signalling that the plants had produced seed and had reached the end of their life. Our timing was perfect. This precious seed is now safely stored at INIA and in the Millennium Seed Bank, UK for future generations.'

**Figure 14.15**
Michael Way collecting seed from the endangered bamboo *Chusquea ciliata*, in Chile.

## Patrick Muthoka

Patrick Muthoka is a plant scientist working for the East African Herbarium based at the National Museum in Kenya's capital city, Nairobi. Patrick's birthplace is in the dry areas of eastern Kenya where he has seen great increases in the pressure on the land because of an ever-growing population. It was this that drove Patrick first to become involved with plant propagation and he pioneered the use of relatively simple 'propagation frames' to help farmers and landowners propagate material from many of the useful tree species found in these arid areas. In 2001, the National Museum established a Plant Conservation Programme and with the help of partners from the National Genebank of Kenya and the Forestry Tree Seed Centre, a project called 'Seeds For Life' was initiated and became a major partner in the MSBP.

Since the beginning of this conservation initiative, Patrick and the Seeds for Life team have been responsible for collecting and conserving seed collections from nearly 2,000 native species in Kenya. Some are from species completely new to science, many represent the first seeds ever seen and certainly the vast majority provide the first opportunity for germination trials to be carried out.

**Figure 14.16**
Patrick Muthoka at work in the field.

Patrick works closely with Kew on an extension project called the 'Useful Plants Project', which is helping farmers and herbalists to propagate the useful plants of their area. This is requiring the delivery of much field-based training in seed collection and storage techniques, nursery establishment, germination and propagation.

The challenge that Patrick and the Seeds For Life team now face is actually re-establishing these rare and useful species back into the wild to help meet Kenya's conservation targets and responsibilities, and ensuring that farmers are able to benefit from the new techniques developed by Patrick and others.

## Fiona Hay

Fiona Hay, another member of the MSBP team, collects and researches seeds from native UK aquatic flowering plant species. In the UK, there are 60–70 native freshwater plant species. Collecting seeds of these species usually involves getting into a dry-suit and wading or snorkelling around ditches, ponds, rivers and lochs. The seeds are then transported, usually in water to prevent them from drying out, back to the MSBP laboratories for preservation.

**Figure 14.17**
Fiona Hay collecting aquatic specimens.

Although we have gone through an age where plants and their products have been extensively exploited, people throughout the world are beginning to realise just how important plants are. Some of the examples in this chapter have shown how trading in plants led to the development of the great botanic gardens of the world. Increasingly, governments and people generally are realising that plants are just too important to ignore and that future trade in them needs to have various safeguards in place in order to control these most valuable of commodities.

# 15 Plants and the future

This concluding chapter touches on some of the threats that plants face in the future, and the consequent challenges to humans who rely on plants for so many things.

## 15.1 From global to local

By the end of the twentieth century, the world seemed a smaller place than it did at the century's start. Now, in the twenty-first century, explorers and plant collectors continue to scour the globe in their hunt for new plants that could act as food sources, building materials, pharmaceuticals, and so on. Plants and their products are global commodities; indeed, many plant products are traded on the world's commodity markets every day.

One only has to look in a supermarket to see fresh fruit and vegetables that have been flown thousands of miles. A new term, 'food miles', aims to make consumers more aware of the distances and therefore the cost in terms of fossil fuels used to transport plant products so far, often between continents. To some, transporting fresh food over long distances is acceptable and they argue that buying such produce provides an income to the farmers and growers in developing countries. However, many are beginning to question the wisdom of creating demand for fresh plant products out of season when it is not possible to source such produce from your own country.

**Figure 15.1 - left**

As this selection of products illustrates, supermarkets are under pressure to show the country of origin of their fruit and vegetables.

**Figure 15.2 - right**

In some countries, the desire to buy locally produced foods and reduce food miles has increased the numbers of farmers' markets, where locally sourced food is sold to local people.

Figure 15.3
Deforestation occurring in the tropical rainforest of the Amazon basin, Brazil, through the use of fire.

The drive to produce ever more food from crops has other effects too. Agriculture has become more mechanised, with a reliance on fossil fuels to power the machinery. In addition, the clearance of habitats such as rainforests to make way for farmland or for land to grow biofuel crops has resulted in an alarming rate of deforestation. In many cases the land is cleared by using fire, something that can now often be monitored by satellite.

However, the pressure to produce more food continues to increase as the world population rises. In addition, in newly affluent countries, such as India and China, there is an increasing expectation of access to more 'exotic' or 'western' foods, such as meat-based products. One of the reasons why the Amazonian rainforests are being cleared is to create more agricultural land to grow soybeans and grain that can be used to feed cattle, and so produce greater quantities of meat.

## 15.2 How might the challenges be met?

The problem of supplying enough food for the world's population is a global issue, and it requires a global solution. For the best part of a decade, many governments have been locked in the world trade talks to try to come up with solutions for how global agricultural markets and trade can function better. It may require some drastic steps to be taken to get resolution and this raises the question: are governments brave enough to take these steps?

As discussed in Chapter 4, growing crops for biofuels can be at the expense of cultivating food crops. In addition, growing plants for food, and indeed biofuels, can be at the expense of maintaining biodiversity (see Chapter 12). These conflicts may be reconciled if some poorer countries were funded by wealthier countries to enable them to maintain a certain level of biodiversity.

One reason that biodiversity is important is for ensuring that plants that are potentially useful to humans do not become endangered or extinct before they can be fully researched. In addition, many wild relatives of crop plants contain useful genes that could be used either in conventional breeding or in genetic modification to increase the productivity of our most important crop plants.

Some countries have realised the importance of looking at all aspects of their policies related to agriculture, both economic and environmental. The European Union countries have abandoned their policy of set-aside (see Chapter 11). This policy had previously benefitted biodiversity but the increasing global demand for more cereals and other crops means that these set-aside areas are being put back into agricultural use. Although some have argued that having more biodiverse farmland is good for the environment, economics and the need to grow more revenue-producing crops becomes paramount.

Governments, with the aim of ensuring food supply to their own populations, fund the majority of agricultural research. Agricultural research can be relatively expensive so private companies undertake only a limited amount of such research. Many developed western countries already allocate a relatively large amount of financial resources to agricultural research, and some other governments are reviewing their funding of basic agricultural research. India has one of the largest public agricultural research systems in terms of workforce in the world, led by the Indian Council of Agricultural Research (ICAR). The ICAR employs more than twice as many scientists as the United States Department of Agriculture. However, if you look at the contributions of these two countries as a percentage of Gross Domestic Product (the amount of money a country makes internally from goods and services), it can be seen that, compared with India, the United States allocates more than ten times the financial resource to agricultural research. It may be an opportune time for countries to look again at the resources they put into plant science.

## 15.3 Plants for the future

At the start of the twenty-first century, the United Nations put forward ten 'Millennium Goals' that they wished the world's governments to work towards, with a target implementation date of 2015. It is telling that they chose 'to halve the proportion of poor and hungry people by 2015' as the goal that should be achieved first. In some parts of the world, this target is getting close to reality. However in other regions, notably sub-Saharan Africa, there has been little movement towards the goal. It perhaps is not a coincidence that while agricultural research expenditure in countries such as India and China has tripled in real terms over the past 20 years, the expenditure in sub-Saharan Africa has increased by barely a fifth and, indeed, the expenditure has declined in half of the countries within this group. Many people think that other more affluent countries should be assisting sub-Saharan African countries to a greater extent in developing their agricultural productivity. The requirement to increase productivity is driven by population growth (see Figure 2.21). In 1850 the world's population stood at around 1,260 million, by 1950 it was 2,520 million and at the time of writing it is more than 6,700 million. The pressure to produce more food, for more people, on the same area of land is ever increasing.

Scarcity of food supply in various parts of the world has become front page news. New methods of cultivating plants may help produce more food, but some argue that new technologies such as micropropagation (see Chapter 9) and even perhaps genetically modified plants (see Chapter 10) are also required. However, in some parts of the world, concerns over using such technologies remain.

But plants are needed for much more than food and other products: they are an essential part of the environment in which we all live. New plant species are continually being discovered in the wild; currently around 2,000 are discovered every year. To give one example, in 2008, scientists from Kew identified a new genus of palms, after a family in Madagascar saw a type of palm they did not recognise. Palms from this new genus, named *Tahina*, are known as 'Madagascan suicide palms' because they die immediately after flowering. It remains to be seen what role this genus plays within its local environment, or indeed what other potential uses this plant may have. There are only a few hundred individuals known in the wild, and if the genus had been allowed to become extinct, then any chance of learning more about it would be gone forever.

**Figure 15.4**

The Madagascan suicide palm (*Tahina spectabilis*). The palm produces a 20-metre high spike of flowers once in its lifetime, after which it dies. The palm is large enough to be visible on satellite imagery, opening up the possibility that other plants new to science may be discovered in a similar way.

Plants also benefit humans in other ways, for instance the cumulative positive effects of working with plants in gardens have been recognised for many centuries. An individual's feeling in a garden can vary from wonder when looking at a garden landscape full of areas to discover, to a sense of security and safety in a more secluded garden setting. All the senses are engaged in a garden: sight, smell, touch, sound, and taste if the garden is growing food. The cyclical and seasonal nature of the work in cultivating a garden and growing plants is said to make people feel closer to nature. In gardens where food is grown, there is the additional reward of eating the fruit and vegetables and the consequent beneficial effect on health. The physical benefits of gardening include better fitness and improved strength, balance and hand–eye coordination. As well as the cyclical nature of gardening, the repetitive nature of many gardening tasks can have a meditative effect, giving time for reflection and relaxation within the work; and gardening is being used increasingly for therapeutic reasons to relieve stress and help people with mental illness. Currently in the UK, many organisations work in gardens or allotments with people with mental illness (including dementia and Alzheimer's disease), brain injury, learning difficulties, visual impairments and more, either offering qualifications or just health benefits and enhanced quality of life and emotional well-being. This is one area in which future work may show additional benefits for humans working with plants.

Although we now know much about plants, how they function and what useful chemicals many of them contain, there is plenty still to be discovered and many opportunities remain for those interested in finding out how plants function within their habitats. For instance, as you saw in Chapter 11, some plants have evolved elaborate communication systems and this was only discovered by researchers within the past decade. Some of the messenger compounds are gases and can signal the chance of an impending insect attack from one plant to another, giving the plant which may be attacked the chance to put into place a range of potential protective mechanisms.

One example which illustrates the interplay between plants and their environment is the process called vernalisation of winter wheat (that is, wheat that is sown in the autumn). Winter wheat plants require a sufficiently long period of cold weather before they will flower and hence set grain the following summer. The extent of climate change has been such that many winter wheat cultivars are now grown at the limit of the shortest required length of cold spell to allow flowering and subsequent grain set to occur. Warmer winters may prevent vernalisation and future winter wheat crops may be much reduced or even fail completely in a changing climate.

As a very late arrival on a planet made habitable for animals by plants, humankind has been completely reliant on plants from the time *Homo sapiens* first evolved. Over hundreds of thousands of years, our understanding of plants has developed at sufficient pace to keep more or less in step with demand for the resources they supply. If we are to deal with the problems of the future (a potential 50% increase in population, climate change, and unexpected and new disasters), we would be wise to retain all our plant inheritance and thereby maximise our capacity to cope.

# Photography credits

All photography ©Andrew McRobb/RBG Kew unless listed below:

**Front cover**
Leaves: ©Robert Kruh.
Money & lentils ©Paul Little/RBG Kew.

**Back cover**
©iStockphoto.com.

**Chapter 1**
Title page ©Paul Little/RBG Kew. Fig. 1.3a The Open University. Fig. 1.4 ©John Durham/Science Photo Library. Fig. 1.6 (plant) ©Siwiak Travel/Alamy.

**Chapter 2**
Fig. 2.4a ©Joe Gough/Dreamstime.com. Fig. 2.4b ©Bob Gibbons/Alamy. Fig. 2.5 ©Science Photo Library. Fig. 2.7a ©Michelle Garrett/www.garden-collection.com. Fig. 2.8a The Metropolitan Museum of Art/Art Resource/Scala, Florence. Fig. 2.8b ©Inga Spence/Alamy. Fig. 2.10 US National Library of Science/Science Photo Library. Fig. 2.11 ©Peter Gau/Dreamstime.com. Fig. 2.18 ©Agripicture Images/Alamy. Fig. 2.19 ©Brian Gadsby/ Science Photo Library. Fig. 2.23 ©Seesea/Dreamstime.com.

**Chapter 3**
Fig. 3.3b & 3.8 ©Peter Gasson. Fig. 3.4 ©eye35.com/Alamy. Fig. 3.5 ©Trevor Hyde/Alamy. Fig 3.6b ©Jim Parkin/Alamy. Fig. 3.7a imagebroker/ Alamy. Fig. 3.10 ©Robert Harding Picture Library Ltd/Alamy. Fig. 3.12a Ethanobotany Collection, RBG Kew. Fig. 3.12b ©Digifoto Gamma/Alamy. Fig 3.12c ©The Biocomposites Centre/Eurelios/ Science Photo Library. Fig. 3.12d ©Eye of Science/Science Photo Library. Fig. 3.13a ©Malkolm Warrington/Science Photo Library. Fig. 3.13b ©Steve Gschmeissner/ Science Photo Library. Fig. 3.14 ©SAV/Alamy. Fig. 3.15 University of Warwick. Fig. 3.17 (cassava) ©Flavio Coelho/Alamy. Fig. 3.17 (rice) ©David Sanger Photography/Alamy. Fig. 3.17 (sorghum) ©Jill Van Doren/Alamy. Fig. 3.17 (yam) ©Bon Appetit/Alamy.

**Chapter 4**
Fig. 4.3a ©James King-Holmes/ Science Photo Library. Fig. 4.3b ©Brian Gadsby/ Science Photo Library. Fig. 4.4 ©Dr Rob Stepney/ Science Photo Library. Fig. 4.5 ©Nigel Cattlin/Alamy. Fig. 4.6 ©Andrew Lawson/www.garden-collection.com. Fig. 4.7 ©Zhang Liwei/Dreamstime.com. Fig. 4.8b ©Immersia/Creative Commons. Fig. 4.10 (wheat) ©George Mastoridis/Dreamstime.com. Fig. 4.10 (sugar beet) ©Wessel Cirkel/ Dreamstime.com. Fig. 4.10 (sugar) ©foodfolio/Alamy. Fig. 4.11©Howard Davies/Alamy.

**Chapter 5**
Fig. 5.1, 5.2 & 5.3b ©Elizabeth Dauncey. Fig. 5.3a ©Robert Bevan-Jones. Fig. 5.5, 5.6, 5.8 & 5.10 ©Hannah Banks. Fig. 5.7 ©Robert Scotland. Fig. 5.11a ©Genevieve Vallee/Alamy. Fig. 5.11b ©Daniel L. Geiger/SNAP/Alamy. Fig. 5.12 ©DNA Landmarks.

**Chapter 6**
Fig. 6.2 Crown Copyright 2007, reproduced with the permission of the Controller of HMSO and the Queen's Printer for Scotland. Fig. 6.4 Collection of the University of Michigan Health System, Gift of Pfizer Inc.. Fig. 6.7 ©Custom Life Science Images/Alamy. Fig. 6.8 ©Real World People/Alamy. Fig. 6.9 ©flowerphotos/Alamy. Fig. 6.10 ©Zenpix/Dreamstime.com. Fig. 6.11a The Open University. Fig. 6.11b ©Nigel Cattlin/Alamy. Fig. 6.12©Catherine Rutherford. Fig. 6.15 ©WILDLIFE GmbH/Alamy. Fig. 6.16 (Pelargonium) ©Derek Harris/The Garden Collection. Fig. 6.16 (Delphinium) ©Steffen Hauser/botanikfoto/Alamy. Fig. 6.17 ©Tielemans/Dreamstime.com. Fig. 6.19 (iris) ©Nicola Stocken Tomkins/The Garden Collection. Fig. 6.19 (currants) ©Jonathan Buckley/www.garden-collection.com.

**Chapter 7**

Fig. 7.1 Bridgeman Art Gallery. Fig. 7.2 ©iStockphoto.com. Fig. 7.3b Ethanobotany collection, RBG Kew. Fig. 7.4a ©Elizabeth Dauncey. Fig 7.4b ©Dr Jeremy Burgess/Science Photo Library. Fig. 7.5 ©Iliuta Goean/Dreamstime.com. Fig. 7.8 ©Inga Spence/Alamy. Fig. 7.9 ©Dr Ernest Hamel, USA National Cancer Institute at Frederick. Fig. 7.10 ©Elizabeth Dauncey. Fig. 7.12 & 7.13 ©Terry Sunderland. Fig. 7.14b ©Emilio Ereza/Alamy.

**Chapter 8**

Fig. 8.1 Artist unknown/RBG Kew. Fig. 8.3 ©MARKA/Alamy. Fig. 8.5 ©Andy Heyward/ Dreamstime.com. Fig. 8.6 ©Chuck Cecil/Science Photo Library. Fig. 8.9 ©Andrew Lawson/ www.garden-collection.com. Fig. 8.10a ©Martin Bond/ Science Photo Library. Fig. 8.10b ©Science Photo Library. Fig. 8.11 ©Enigma/Alamy.

**Chapter 9**

Fig. 9.1 ©Jonathan Buckley/www.garden-collection.com. Fig. 9.5, 9.7 & 9.8a ©V Sarasan. Fig. 9.6 ©Swiss Orchid Foundation at the Herbarium Jany Renz (http://orchid.unibas.ch). Fig. 9.8b ©Lucy Hart. Fig. 9.11 ©W. J. Baker. Fig. 9.12 ©John Dransfield. Fig. 9.13 The Open University.

**Chapter 10**

Title page ©Alexander Gitlits/Dreamstime.com. Fig. 10.2 © Steve Percival/Science Photo Library. Fig. 10.4 ©Andrew Dunn. Fig. 10.5a &b ©James King-Holmes/Science Photo Library. Fig. 10.6a & c ©Chris Clennett. Fig. 10.7 Iowa State University. Fig. 10.8 ©Leonard Lessin/Science Photo Library. Fig. 10.11©Bram Van Loock and Prof. Kris Vissenberg, University of Antwerp. Fig. 10.12 ©Pioneer Hi-Bred. Fig. 10.13 ©Bill Barksdale/Science Photo Library. Figs. 10.14, 10.15 & 10.16 ©PBARC. Fig. 10.17 ©Courtesy Golden Rice Humanitarian Board (www.goldenrice.org).

**Chapter 11**

Title page ©Paul Little/RBG Kew. Fig. 11.1a ©RF Company/Alamy. Fig. 11.1b & 11.2b ©Nigel Cattlin/Alamy. Fig. 11.2a ©Catherine Rutherford. Fig. 11.2c ©Graphic Science/Alamy. Fig. 11.3 (goldenrod) ©Antje Schulte/Alamy. Fig. 11.3 (Michaelmas daisy) ©Anthony Collins/Alamy. Fig. 11.5a (adult ladybird) ©Andrew Darrington/Alamy. Fig. 11.5 (larvae) ©blickwinkel/ Alamy. Fig. 11.7 (stemborer) ©Marlin E. Rice/Science Photo Library. Fig. 11.7 (*Striga* weed) ©Institute for Genomic Diversity at Cornell University, USA. Fig. 11.8 ©*icipe* and Rothamsted Research. Fig. 11.9 ©Simon Jasperse, Kilmo/FAO Plant Nutrition Programme. Fig. 11.10 ©Emilio Ereza/Alamy. Fig. 11.13 ©Thérèse Arcand. Fig. 11.14a ©Molly Welsh. Fig. 11.14b & 11.14c ©Nigel Cattlin/Alamy. Fig. 11.15b ©Liquidphoto/Dreamstime.com.

**Chapter 12**

Fig. 12.2a ©Gary Crabbe/Alamy. Fig. 12.2b ©John Game. Fig. 12.3 ©Andrew Lawson/The Garden Collection. Fig. 12.4 ©Sue Wickison. Fig. 12.5 ©Muséum National d'Histoire Naturelle, Paris. Fig. 12.6 ©Fabian A. Michelangeli. Fig. 12.7a www.plantlife.org.uk. Fig. 12.7b & c Bob Gibbons/Science Photo Library. Fig. 12.8a Bob Gibbons/Alamy. Fig. 12.9 ©Jerl71/Dreamstime.com. Fig. 12.10 ©Andrew Lawson/The Garden Collection. Fig. 12.11 ©George Schatz. Fig. 12.12 ©Forest & Kim Starr. Fig. 12.16 ©Chris Hellier/Science Photo Library.

**Chapter 13**

Fig. 13.1 ©CBD. Fig. 13.2 ©Dr Jeremy Burgess/Science Photo Library. Fig. 13.3 ©Bill Heinsohn/Alamy. Fig. 13.4 ©David Hosking/Alamy. Fig. 13.5 ©Brett Baunton/Alamy. Fig. 13.6 ©Borderlands/Alamy. Fig. 13.7 ©ParoliGalperti/ Cuboimages. Fig. 13.8 ©Liz Eddison/The Garden Collection. Fig. 13.12 ©Rebecca Cairns-Wicks. Fig. 13.13 © Colin Clubbe. Fig. 13.14 ©Chris Hellier/Science Photo Library.

**Chapter 14**

Fig. 14.1a ©Ashmolean Museum, University of Oxford, UK/The Bridgeman Art Library. Fig. 14.1b ©Weldon Schloneger/Dreamstime.com. Fig. 14.2b Lincolnshire County Council (THE COLLECTION, Art and Archaeology in Lincolnshire). Fig. 14.3a ©Kew Collection. Fig. 14.3b ©Catherine Rutherford. Fig. 14.5 ©Milestone Media/Alamy. Fig. 14.6 ©Protea Atlas Project. Fig. 14.7a&b ©Phillip Cribb. Fig. 14.8 ©Irina Ponomarenko/Dreamstime.com. Fig. 14.9 ©Nico Smit/Dreamstime.com. Fig. 14.10 ©Liane Matrisch/Dreamstime.com. Fig. 14.13 RBG Kew. Fig. 14.14 & 14.15 ©John Dori. Fig. 14.16 Tim Pearce/RBG Kew.

**Chapter 15**

Fig. 15.2 ©www.shabdenparkfarm.com. Fig. 15.3 ©Les Gibbon/Alamy. Fig. 15.4 ©John Dransfield.

Every effort has been made to contact and accurately credit all copyright holders. If we have been unsuccessful, we apologise and welcome correction for future editions and reprints.

# Acknowledgements

The following people have made valuable contributions to this book:

**Project and text development**
Hannah Banks, Colin Clubbe, Angela Colling, Felix Forest, Gina Fullerlove, Carol Furness, Pat Griggs, Pat Heslop-Harrison, Sue Hunt, Rogier de Kok, Chris Leon, David Mabberley, Angela McFarlane, Noel McGough, Isla McTaggart, Mark Nesbitt, Christine Newton, Alan Paton, Tim Pearce, Jill Preston, Margaret Ramsay, Viswambharan Sarasan, Monique Simmonds, David Simpson, Roger Smith, Nigel Taylor, China Williams and Michael Way

**Picture research**
Fiona Bradley, Liz Eddison, John Harris, Andrew McRobb, Justin Moat and Catherine Rutherford

**Project management, copyediting and proof reading**
Michelle Payne, Sharon Whitehead

**Production**
Lloyd Kirton

**Design and layout**
Nicola Thompson, Culver Design

**Cover design**
Lyn Davies

# Index


abaca, 35, 36
acetylsalicylic acid, 78
aconitine, 52
Adam's mistletoe, 139
African cherry, 80, 82, 83, 156
African oil palm, 109
agricultural research, 175
agriculture, 13, 14
alcohol, effect of, 97
alcoholic fermentation, 96
alder, 68
alfalfa, 125, 132
alien invasive species, 138, 140–2, 147, 158
alkaloid, 77, 92
allelopathy, 132
alliums, 70
α-linolenic acid, 23, 24, 66
amylose, 37
annual growth rings, 30, 55
anther, 58
anthocyanins, 72, 73
antioxidants, 65, 73, 91
  what are they ? 65
aromas, 68–70
Asian slipper orchid, 165
aspirin, 77–9
atropine, 77
Australian clover, 134
Australian pine, 159
Australian swamp stonecrop, 141
authentication of medicinal herbs, 86, 87
autotrophic, 63
auxin, 102, 117
Ayurvedic medical systems, 76

bamboo, 170
bananas, 108, 109
Banks, Joseph, 162
bark, 30, 33, 77, 82, 83
bastard cabbage, 135
bastard gumwood, 157
beans, 20, 21, 68, 134
beer, 96
benign prostatic hyperplasia (BPH), 83
bergamot, 90
beriberi, 20
β-carotene, 71, 123
big-leaved mahogany, 32, 168
biodiesel, 46, 47
biodiversity, 49, 106, 130, 137, 138, 141, 145, 146, 149, 153, 164, 174, 175
  definition of, 137
  hotspots, 106, 145–7, 169
  in relation to set-aside, 129, 130, 175
  loss of, 138–43
  threats to, 139–145
bioethanol, 46, 47, 49
biofuels, 24, 40–9
  biofuel crops, 174
  for transport, 46
  grasses as an energy source, 44
  rape oil as a biofuel, 21
  some of the issues, 48, 49
biofumigation, 135
biogas, 48, 49
biological control, 127
biomass, 41, 44, 49
biopiracy, 168
bioplastics, 39
black walnut, 133
blackberries, 73
blackcurrants, 73
blueberries, 73
botanic gardens, 154, 155
bottle palm, 108
BPH — see benign prostatic hyperplasia
brassica family, 24, 70
broccoli, 70
brown galingale, 140
Bruno Richard Hauptman, 54, 55
brussel sprouts, 70
Bt toxin, 120, 121, 124

cabbage, 70
cactus moth, 141
caffeine, 91, 92, 93
cannabis, 97–99
canola, 23, 125
carbohydrates, 19, 20, 63, 66
carbon, 8
carbon cycle, 42
carbon footprint, 43
carbon neutral, 43
carbon sequestration, 43
carnivorous plants, 103, 167
carotenes, 71, 123
carotenoids, 68, 71
carpel, 58
carrot root fly, 127
cassava, 38
castor oil plant, 53
catechins, 91
CBD, 149–151, 154, 168, 169
cells, 7, 29, 102, 117
cellulose, 7, 54, 66
centres of origin, 17
cereals, 19
chives, 70
chlorophyll, 8, 9, 71
  role in the plant, 8
  green colour of, 9, 71
chloroplasts, 7, 8
chocolate, 22, 94, 95
chromosomes, 112, 117, 118
cinnamon, 68, 90
CITES, 139, 164–8
citronella, 68, 69
citrus fruit, 65
classification, 10, 11
climate change, 27, 111, 138, 143, 177
clone, 61, 101
cloves, 68, 90
coca leaves, 93
cocaine, 93
cocoa, 94, 95
coconut palm, 34
coffee leaf rust disease, 91
cola, 93
colour in plants, 71–3
  why are plants green? 9
communities of plants, 54
companion plants, 130–2
  as shelters, 132
  to control weeds, 131
Conference of the Parties (COP), 149
Conservation Biotechnology Unit, Kew, 106
  success with orchids at, 106
conservation, 84, 149, 150
  *ex situ*, 150, 154, 155, 164
    definition of, 150, 154
    categories of, 154
  *in situ*, 84, 150
    definition of, 150
    of African cherry, 84
  of island habitats, 157
Conservation International, 145
Convention on Biological Diversity — see CBD

Convention on International Trade in Endangered Species — see CITES
coppice, 43
  short rotation, 44
cork, 33, 34
  cork oak, 33
  TCA and the tainting of wine, 34
cornflower, 72
corrective growth, 55
cotton, 34, 35, 121, 124, 125
cotyledons, 14, 21, 94
critically endangered species, 149
crop improvement, 14
crops, 13, 111
  first food crops, 13
  for oil production, 21
cross-fertilisation, 15, 18
croton tree, 46
cultivars, 113, 115
cutin, 66
cuttings, 101, 113
cyanidin, 72
cycad, 166
cyclamen, 166
cytokinin, 102

dandelions, 56
date palm, 107
DDT insecticide, 127
deadly nightshade, 77
decaffeinated coffee/tea, 92
defence mechanisms in plants, 133–5
deforestation, 49, 174
delphinidin, 72
delphinium, 72
dicot, 14
dicotyledonous, 14
digoxin, 52
dioecious, 142
DNA, 60, 61, 112, 115, 118, 142
  analysis of Japanese knotweed, 142
  bases within, 112
  databases, 61
  profile, 61
  transferred (t-DNA), 117
  use in forensic science, 60
  within genes, 112

Ebers papyrus, 75
ecosystems, 54, 127, 138, 150
Edmond Locard, 54
embryo, 14, 19, 20, 108
*Encarsia* wasps, 128
endangered species, 144, 149
endemic species, 140, 145, 146, 157
endosperm, 14
energy content of plant products, 41
enzymes, 96
epicotyl, 14
essential fatty acids, 23, 24
ethnopharmacology, 82
ethylene, 134
European larch, 114
European yew, 51, 81, 82
explants, 102
extinctions, 137, 138, 144, 146, 164, 176
  recording, 143

Fairtrade, 93
fats, 66, 67
fatty acids, 22–4
fennel, 129
fermentation, 20, 96
Fertile Crescent, 13
fertilisers, 25
  role in boosting crop production, 25
  role in nitrogen pollution, 143
fever bark tree, 76–8, 163
fibres, 29, 34–7
  in ropes, 35
  tissue present in, 29
field gene banks, 154, 157
filament, 58
Fiona Hay, 171
flavanols, 72, 73, 91
flavones, 72, 73
flavonoids, 72, 73, 93
flavours, 68–70
flax, 23, 35
flowers, 6, 71, 114, 177
  basic function of, 4
  colours of, 71–3
  structure of, 58
food chain, 5, 63, 127
food miles, 173
'food of the gods', 94
food security, 25
forensic science, 54, 55, 58–60
Forest Stewardship Council (FSC), 33
Formula 3 green racing car, 37
fossil fuels, 41–3, 48, 49
foxglove, 52
free radicals, 65
fruit, 64–66, 71, 91, 95, 173
  as a source of vitamin C, 64, 65
  coffee, 91, 92
  colour of, 71, 73
  within a diet, 63, 64
fumitory, 57
fungal disease, 109
fungal mycorrhiza, 104, 150

gall tumours, 116
garlic, 70
Gaucher disease, 124
gene transfer, 118
genes, 112, 114, 119, 139
  virulence, 117
genetic modification, 116–9
  by viruses, 119
genetically modified (GM) plants, 111
  crops, 123, 124
  examples of, 119–24
  in the pharmaceutical industry, 124
  techniques for producing, 116–9
  with herbicide resistance, 120
  with insect resistance, 120, 121
  with virus resistance, 121
  why might they be useful? 115
genome, 112, 117, 119, 139
genus, 10, 11, 114
Georgi Markov, 53
geranium, 56
germination, 14, 15, 104, 105
  symbiotic, 104, 105
giant redwoods, 135
ginger, 68, 90
global warming, 42, 143
glucose, 37, 38, 66
glucosinolates, 24, 70, 133–5
glycerol, 22
glyphosate, 120
GM plants — see genetically modified plants
Golden Rice, 123
goldenrod, 129
grains, 14, 19
  as a food crop, 19
  production of, 111
grapes, black, 73
grasses, 56
  as an energy source, 44, 45, 48, 49
green manure, 135

Green Revolution, 19
greenhouse gas, 42, 43, 49, 143
Gregor Mendel, 113
growth rings, 30, 55

habitat transformation, 138, 140, 147
habitats, 54, 139, 150
harebell, 10, 59
hazel, 56
heartwood, 30
heathers, 56
hedge woundwort, 55
hemp, 35, 36, 97
    manilla hemp, 36
herbal treatments, 76, 77
herbaria, 155
herbicides, 127, 132
heterotrophic, 63
HFCS — see high fructose corn syrup
Hickman's potentilla, 138
high fructose corn syrup (HFCS), 38
hogweed, 55
holly, 113
Hooker, Joseph Dalton, 162, 163
hops, 96
horse chestnut, 161
hotspots, 106
    definition of, 145
    for orchids, 106
    Madagascar, 146, 147
hoverflies, 129
human diet, 63–8
human migration, 137
Humanitarian Use Licenses, 123
hunter-gatherers, 13, 138
hybrids, 113, 114
hypocotyl, 14

inkberry, 158
insect repellents, 68, 69
intercropping, 132
International Maize and Wheat
    Improvement Centre (CIMMYT), 27
International Union for the Conservation
    of Nature (IUCN), 108, 144, 149,
    151
invasive alien species, 138, 140–2, 147,
    158
island habitats, 157–9
isoflavones, 134
isoprene unit, 68
isoprenoids, 68

Japanese knotweed, 142
Japanese larch, 114
jasmine, 90
jute, 34

kola nuts, 93
kudzu, 158

ladybird, 128, 130
ladybug, 128, 130
lady's slipper orchid, 106
land management agreements, 153
larch, hybrid, 114
latex, 163
leaves, 6, 51, 71, 90, 97
    basic function of, 4
    containing poisons, 51
    colour of, 9, 71
    in tea production, 90
    in cannabis production, 97
legumes, 20, 21
lemon balm, 70
lemon, 68
lemongrass, 69
Leyland cypress, 114
lignin, 29, 30, 31, 36, 54, 72
lime, 32
linden, 32
linoleic acid, 23, 24
lipids, 66, 67
lobelias, 144
lungwort, 76
lutein, 71
lycopene, 71

macronutrients, 63
Madagascan suicide palm, 176
Madagascar periwinkle, 79
Madagascar, 146, 147
maize, 19, 38, 39, 61, 125, 131
Malagasy palm, 146
mallow, 56
maple, 57
marigolds, 130
medicinal plants, 75–87
Mediterranean corn borer, 120
methyl jasmonate, 134
methyl salicylate, 134
Michael Way, 169, 170
Michaelmas daisy, 129
micronutrients, 63–5
micropropagation, 84, 101-9, 118
    basic process of, 101
    for generating a genetically modified
        plant, 118
    how it works, 102
    of African cherry, 84
    of high-value plants in agriculture,
        108
    of orchids, 104
    of palms, 107, 108
    uses of, 103
Millenium Goals, 175
Millennium Seed Bank Project (MSBP),
    156, 169–71
mint, 68
monkshood, 52
monocot, 14
monocotyledonous, 14
monomer, 37
monoterpenoids, 68, 133
Monsanto, 125
Monterey cypress, 114
Monterey pine, 138
morphine, 76, 77
mustard, 70
mustard greens, 135
mutation, 112, 113
mycorrhiza, 104, 150

naming plants, 10, 11
Napier grass, 131
nasturtiums, 130
national nature reserves, 151, 152
national parks, 151
naturally occurring predators, 128–30
    planting to encourage, 129
Neolithic period, 31
nettles, 55, 129
nicotine, 77
Nikolai Vavilov, 17
nitrate, 67
nitrogen, 20, 67, 143
nitrogen cycle, 67
nitrogen fixation, 20, 67, 68
nitrogen pollution, 138, 143
non-staple food crops, 21–4
Nootka cypress, 114
Norway spruce, 133
nucleus, 7, 112, 117

basic function of, 7
containing the genes, 112

oil crops, 21–4, 109
oil palm, 41, 109
oils, 21–4
saturated oil, 22
unsaturated oil, 22–4
vegetable oils, 22
oilseed rape, 23, 24, 46, 125
oleander, 52
onions, 70
opium poppy, 76, 77
orange, 68
orchids, 104–6, 147, 165
oriental plane tree, 44
ornamentals, over-exploitation due to trade of, 165
ovary, 58
over-exploitation, 139
due to trade, 165
ovule, 58
ox-eye daisies, 129
oxidative fermentation, 90

Pacific yew tree, 80–2
ecological treats to 81
palm oil, 46, 109
palms, 107, 108, 176
Palo verde tree, 60, 61
palynology, 55–7, 59
forensic palynology, 55
papaya, 122, 123, 125
papaya ringspot virus (PRSV), 122, 123
Paris daisy, 116
parrot's feather, 140
particle bombardment, 118, 119, 122
Patrick Muthoka, 170
peas, 20, 113
pelargonidin, 72
pelargonium, 72
peppermint, 90
pesticides, 120, 127, 131,132
pests, 127, 128
naturally occurring predators of, 127, 129
petals, 58
pharmaceutical drugs, 75, 79
phenotype, 112
phloem, 29, 30, 34, 83, 131
phosphate pollution, 138, 143, 147
photosynthesis, 6, 8, 19, 37, 41, 43, 66
basic process of, 8
in biofuel production, 41
in providing energy through carbohydrates, 66
in starch production, 37
role in semi-dwarf varieties, 19
phyto-remediation, 44
phytosterols, 124, 134
pine tortoise scale, 158
pitcher plants, 103, 167
PLA plastic, 39
plant breeding, 15, 18, 112, 113
basic process of, 15, 16
history of, 112
limitations of traditional methods, 115
modern process, 18
selective, 15, 16, 114
plant collecting, 161–71
early plant collectors, 161–3
in the present day, 169–71
plant fibres, 34–7
plant growth regulators, 102, 117, 134
plant names, 10, 86
plant oils, 22–4
health benefits of, 24
plantains, 57
plasmids, 117, 118
poisons, 51–3, 77
pollen, 55–9, 114
animal dispersal of, 56, 57
of pine species, 56, 58, 59
pollen rain, 56–9
pollen structure, 56, 57
wind dispersal of, 57, 58
polylactic acid plastic, 39
polymer, 37, 66
polyphenols, 91
poplar, 44
prickly pear, 141
proteins, 21, 63, 67, 112, 124
as macronutrients, 63
in legumes, 21
role of genes in the production of, 112
proxy indicators, 54
pulses, 19, 20, 21
push-pull planting, 131

quinine, 76, 163
radicle, 14
ramin, 168
red cabbage, 73
Red Lists (IUCN), 149
respiration, 66
rhizomes, 44, 45, 142
rhubarb, 51
rice, 19, 26, 38, 123
as a genetically modified crop, 123
as a source of starch, 38
as a staple food, 19
increasing price of, 26
ricin, 53
Rio Earth Summit, 149
rocket, 135
root nodules, 67
roots, 6, 67, 68
basic function of, 6
nodules and nitrogen fixation, 67, 68
ropes, 35
rosemary, 70
Roundup Ready ®, 120
Royal Botanic Gardens, Kew, 14, 15, 155
authentication of traditional Chinese medicines, 14, 15
herbarium of, 155
micropropagation of the bottle palm, 108
work on tackling alien invasive species, 158
rubber tree, 163

sacred lotus, 162
sage, 70
salicin, 78
salicylic acid, 78
sapwood, 30
scarlet bottlebrush, 132
sclereids, 29
sclerenchyma, 29
scurvy, 65
sea hog's fennel, 152
seeds, 6, 14, 15, 19, 51, 91, 94, 104, 144
as staple foods, 19
basic structure of, 14
conservation, 156, 157, 169–71
containing poisons, 51, 53
dormancy, 157
in the conservation of threatened species, 144, 156
of cocoa, 94

of coffee, 91
of orchids, 104
role in plant selection, 15
seed banks, 103, 156, 157
selection, 15
selective breeding, 15
sepal, 58
sesquiterpenoids, 68, 133
set-aside, 129, 130, 175
shelter belts, 132
silver leaf desmodium, 131
sisal, 34
site of special scientific interest (SSSI), 153
Sitka spruce, 133
snowdrops, 73, 139
Soham murders, 55
somatic embryo, 108
sorghum, 38
soybean oil, 46
soybean, 41, 112, 120, 124, 125
Spanish sage, 70
spearmint, 69
species, 10, 11
spiderwort, 161
spores, 51, 54, 58, 59
sporopollenin, 54
squash, 125
stamen, 58
staple food crops, 19–21
star cactus, 164
star fruit, 141
starch, 19, 37–9, 63
starch crops, 29, 37–9
uses of, 38
stem borers, 131
stems, 6, 29
stigma, 58, 114
strawberry, 101
striga weed, 131
striking a cutting, 101
suberin, 30, 33, 66
sugar beet, 17, 41, 47, 125
sugar cane, 17, 41, 47, 49
sugars, 8, 9, 10, 37, 46, 47, 96, 102
annual production of, 9
basic role of, 8
in alcohol production, 96
in bioethanol production, 46, 47
in micropropagation, 102
in starch production, 37
production by agave, 10
sundews, 103
sunflower oil, 46
sunflowers, 21, 46
sustainability, 149
switch grass, 45, 48, 49
symbiosis, 104, 105

tannins, 72, 73, 75, 133
taxol, 80–2
discovery of, 80
effect on animal cells, 80, 81
TCA (trichloroanisole), 34
tea, 89, 90
chemistry of, 91
terpenoids, 68
testa, 14, 20
tetrahydrocannabinol (THC), 98
theobromine, 93, 95
theophylline, 91
thistles, 56
Ti plasmid, 117, 118
tissue culture — see micropropagation
tobacco mosaic virus (TMV), 121, 122
tobacco, 77, 115
tomatoes, 71
totipotency, 101
trace evidence, 54
trade, 165, 167, 168
of ornamentals, 165
of plants for medicinal use, 167
of wood products, 168
Tradescants, 161
traditional Chinese medicine, 76, 85–7
authentication of, 86, 87
globalisation of, 85
tree ferns, 166
tree of heaven, 133
trichloroanisole (TCA), 34
tropical eaglewood tree, 87
tumour-inducing plasmid, 117

unsaturated oils, 22–4

vegetables, 66, 70, 173
Venus flytrap, 167
virulence genes, 117
virus, 115
virus cross-protection, 121–3
vitamins, 19, 20, 63–5
as micronutrients, 63–5
fat-soluble, 64
presence in grains, 19
vitamin A, 64, 71, 123
deficiency, 123
vitamin B, 64
vitamin C, 64, 65
vitamin E, 24, 64
water-soluble, 64
vulnerable species, 149

walnut, 116, 133
western red cedar, 32
wheat, 16, 19, 38, 47
as a source of starch, 38
as a staple food, 19
in bioethanol production, 47
origin of domesticated wheat, 16
white clover, 59
white currants, 73
white iris, 73
white pine weevil, 133
whitefly, 128
white-topped pitcher plant, 167
wilderness areas, 152
willow, 44, 56, 78
wolf's bane, 52
wood, 29–33
as an energy source, 43
uses of, 31, 41
world population, 25
meeting the challenges of, 25, 174, 175
world trade talks, 174

xanthophylls, 71
xylem, 29, 31, 54, 83

Yadon's piperia, 138
yam, 38

zeaxanthin, 71
zero carbon footprint, 43

# Index of scientific names

Acanthaceae, 57
*Acer*, 57, 72
*Aconitum napellus*, 52
*Aesculus hippocastanum*, 161
*Agave sisalana*, 34
*Agave tequilana*, 10
*Agrobacterium tumefaciens*, 116, 118, 119
*Ailanthus altissima*, 133
*Anacamptis morio*, 105
*Aquilaria sinensis*, 87
*Argyranthemum fructescens*, 116
*Aristolochia*, 86
*Aristolochia manshuriensis*, 86
*Aster*, 129
*Astrophytum asterias*, 164
*Atropa belladonna*, 77
*Bacillus thuringiensis*, 120
*Brassica juncea*, 135
*Brassica napus*, 23
*Bulbophyllum elliottii*, 147
*Cactoblastis cactorum*, 141
*Callistemon citrinus*, 132
*Camellia sinensis* var. *assamica*, 89
*Camellia sinensis* var. *sinesis*, 89
*Campanula rotundifolia*, 59
*Cannabis sativa* subsp. *indica*, 97, 98
*Cannabis sativa* subsp. *sativa*, 97, 98
*Casuarina equisetifolia*, 159
*Catharanthus roseus*, 79
*Cedrela odorata*, 168
*Chusquea ciliata*, 170
*Cibotium*, 166
*Cinchona*, 163
*Cinchona calisaya*, 76, 77, 78
*Cirsium*, 56
*Citrus* x *limon*, 90
*Cocos nucifera*, 34
*Coffea arabica*, 91
*Coffea canephora*, 91
*Cola acuminata*, 93
*Cola nitida*, 93
*Commidendrum rotundifolium*, 157
*Corchorus*, 34
*Corylus*, 56, 59
*Crassula helmsii*, 141
*Crinodendron hookerianum*, 163
*Croton megalocarpus*, 46
*Cupressus macrocarpa*, 114
*Cuprocyparis leylandii*, 114
*Cyathea*, 166
*Cyclamen persicum*, 166
*Cylindrocline lorencei*, 104
*Cymbopogon*, 69
*Cyperus fuscus*, 140
*Cypripedium calceolus*, 105
*Damasonium alisma*, 141
*Delonix leuchantha*, 56
*Desmodium uncinatum*, 131
*Dicksonia*, 166
*Digitalis purpurea*, 52
*Dinizia excelsa*, 56
*Dionaea*, 167
*Dionaea muscipula*, 167
*Drosera*, 103
*Dypsis ambositrae*, 146
*Elaeis guineensis*, 109
*Encarsia formosa*, 128
*Encephalartos woodii*, 166
*Eperua*, 57
*Ephedra andina*, 169
*Eruca sativa*, 135
*Erythroxylum coca*, 93
*Fallopia japonica*, 142
*Foeniculum vulgare*, 129
*Fragaria*, 101
*Fumaria*, 57
*Galanthus*, 139
*Galium aparine*, 10
*Gilbertiodendron bilineatum*, 56
*Glycine max*, 124
*Gonystylus*, 168
*Gossypium hirsutum*, 34, 121, 124
*Helianthus annuus*, 21
*Heracleum sphondylium*, 55
*Hevea brasiliensis*, 163
*Hoodia gordonii*, 84, 167
*Humulus lupulus*, 96
*Hyophorbe amaricaulis*, 108
*Hyophorbe lagenicaulis*, 108
*Jatropha curcas*, 46
*Juglans nigra*, 133
*Larix decidua*, 114
*Larix kaempferi*, 114
*Larix* x *marschlinsii*, 114
*Leucanthemum vulgare*, 129
*Malva sylvestris*, 56

*Medicago sativa*, 132
*Melissa officinalis*, 70
*Mentha spicata*, 69
*Miscanthus* x *giganteus*, 45
*Musa*, 109
*Musa acuminata*, 108
*Musa balbisiana*, 108
*Musa textilis*, 35, 36
*Myriophyllum aquaticum*, 140
*Nelumbo nucifera*, 162
*Nepenthes*, 103, 167
*Nerium oleander*, 52
*Nicotiana tabacum*, 77
*Olea europaea*, 22
*Opuntia*, 141
*Oryza sativa*, 123
*Panicum virgatum*, 45
*Papaver somniferum*, 77
*Paphiopedilum rothschildianum*, 165
*Parksonia*, 60
*Pennisetum purpureum*, 131
*Pericopsis elata*, 32
*Peucedanum officinale*, 152
*Phalaenopsis*, 165
*Phalaris canariensis*, 45
*Phaseolus lunatus*, 134
*Phaseolus vulgaris*, 68
*Phoenix dactylifera*, 107
*Phytoseiulus persimilis*, 134
*Picea abies*, 133
*Picea sitchensis*, 133
*Pinus*, 59
*Pinus caribaea* var. *bahamensis*, 158
*Pinus radiata*, 138
*Piperia yadonii*, 138
*Pissodes strobi*, 133
*Plantago*, 57
*Platanus orientalis*, 44
*Plazia cheiranthifolia*, 170
*Poa pratensis*, 57
*Populus*, 44
*Potentilla hickmanii*, 138
*Protea odorata*, 164
*Prunus*, 57
*Prunus africana*, 82, 156
*Psila rosae*, 127
*Pueraria lobata*, 158
*Pulmonaria officinalis*, 76
*Quercus suber*, 33
*Rapistrum rugosum*, 135
*Rheum rhaponticum*, 51
*Rhipsalis*, 166
*Rhizobium*, 20
*Ricinus communis*, 53
*Rosmarinus officinalis*, 70
*Salix*, 44, 78
*Salvia lavandulifolia*, 70
*Salvia officinalis*, 70
*Sarracenia*, 167
*Sarracenia leucophylla*, 167
*Scaevola plumieri*, 158
*Scaevola taccada*, 158
*Sequoiadendron giganteum*, 135
*Sesamia nonagrioides*, 120
*Solidago*, 129
*Stachys silvatica*, 55
*Stemona japonica*, 86
*Stemona sessilifolia*, 86
*Stemona tuberosa*, 86
*Sterculia ceramica*, 140
*Sterculia khasiana*, 140
*Striga hermonthica*, 131
*Swietenia macrophylla*, 32, 168
*Tagetes*, 130
*Tahina spectabilis*, 176
*Takhtajania perrieri*, 144
*Taraxacum*, 56
*Taxus baccata*, 51, 80, 81
*Tetranychus urticae*, 134
*Theobroma cacao*, 94
*Thuja plicata*, 32
*Tilia*, 32
*Toumeyella parvicornis*, 158
*Tradescantia*, 161
*Trifolium repens*, 59
*Trifolium subterranean*, 134
*Trilepidea adamsii*, 139
*Triticum*, 10
*Tropaeolum majus*, 130
*Urtica dioica*, 55, 129
*Viola cryana*, 140
*Xanthocyparis nootkatensis*, 114
*Zea mays*, 38, 39, 61, 131